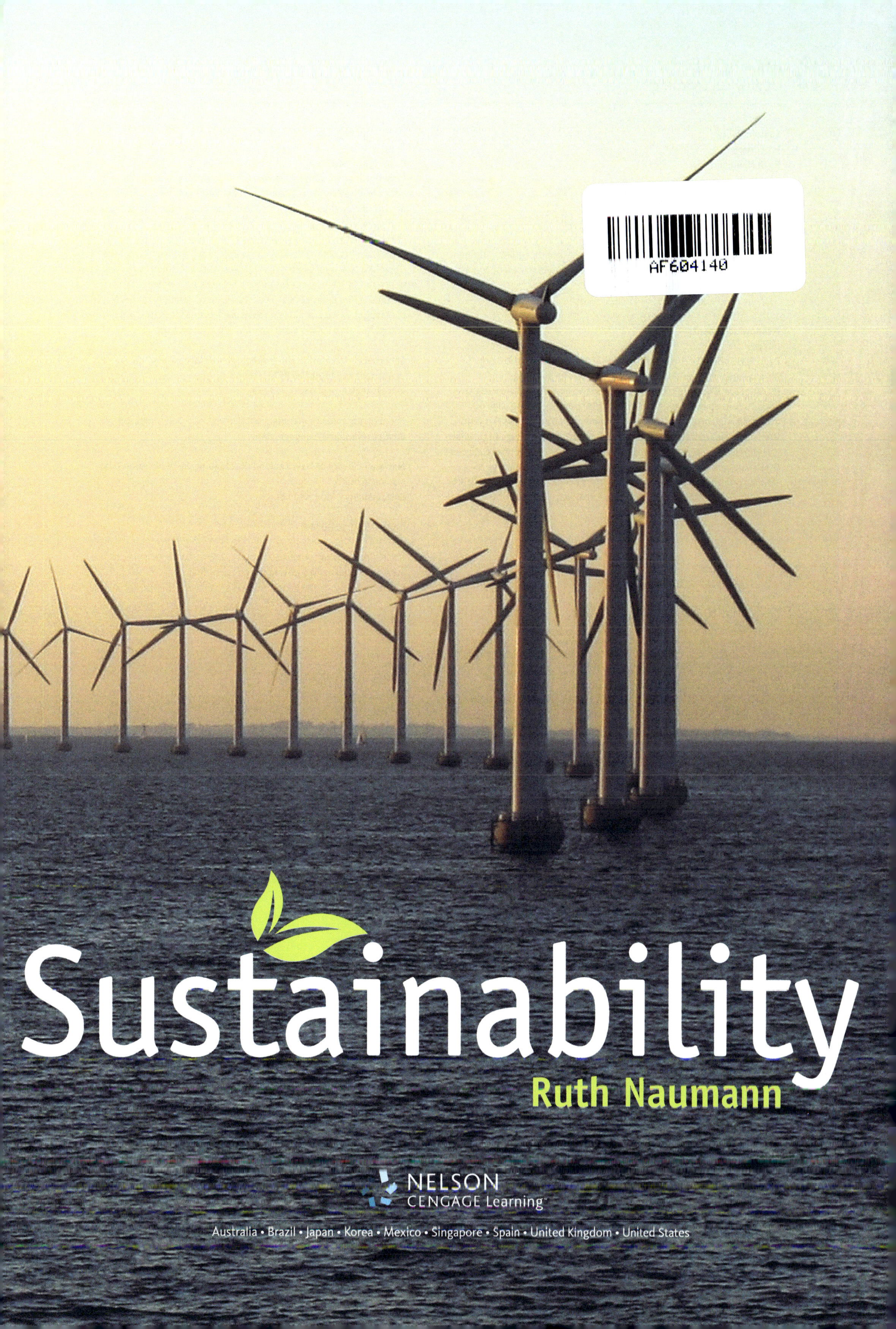

Sustainability

Ruth Naumann

Australia • Brazil • Japan • Korea • Mexico • Singapore • Spain • United Kingdom • United States

Sustainability
1st Edition
Ruth Naumann

Text and cover design: Cheryl Rowe
Cartoon artwork: Laurence Clark
Production controller: Siew Han Ong & Jess Lovell

Any URLs contained in this publication were checked for currency during the production process. Note, however, that the publisher cannot vouch for the ongoing currency of URLs.

Acknowledgements
Images
Page 12 US Antarctic Program; page 15 (Tarawa) and page 17 350.org (Aaron Packard); page 21 (ETS diagram) Ministry for the Environment (2007); page 27 (top), 33 (top), 80, 105 (top two images) Richard Redgrove; page 29 Shweeb Holdings; page 35, 51 NASA; page 50 SCION; page 60 Whangamarino map produced by Environment Waikato for the National Wetland Trust, from the booklet Our Wet and Wild Places; page 61 (Rachel Carson) USFWS; page 64 Icebreaker (photographer Jono Rotman); page 90 Department of Conservation (photographer Dave Alcock); page 95 (Mining protest) Royal Forest and Bird Protection Society of New Zealand; page 96 USAID in Africa; page 98 Green Cabs. Shutterstock for other images.

National Library of New Zealand Cataloguing-in-Publication Data
Naumann, Ruth.
Sustainability / Ruth Naumann.

ISBN 978-017021-034-8

1. Sustainable living—Juvenile literature. 2. Environmentalism—Juvenile literature. [1. Sustainable living. 2. Environmentalism.]
I. Title.

333.72—dc 22

Cengage Learning Australia
Level 7, 80 Dorcas Street
South Melbourne, Victoria Australia 3205

Cengage Learning New Zealand
Unit 4B Rosedale Office Park
331 Rosedale Road, Albany, North Shore 0632, NZ

For learning solutions, visit **cengage.com.au**

Printed in Australia by Ligare Pty Limited.
4 5 6 7 8 9 10 21 20 19 18 17

Contents

What Does Sustainability Mean?

We sustain our world because we hold it up and don't let it fall down and break. We look after it. We don't waste things. We even collect urine from public toilets. It has ammonia in it. So it's good for cleaning. And it makes our togas extra-white.

A lot of English words come from Latin. It was the language that Ancient Romans spoke. 'Sustain' (and 'sustainable' and 'sustainability') comes from the Latin 'sustinere'. 'Tenere' means 'to hold' and 'sus' means 'up'.

Sustainability is a global thing. Global means the whole globe, or the whole world, or the whole planet, or the whole Earth, or the whole Planet Earth.

Sustainability is about the environment. Your environment is what surrounds you. Your natural environment contains things like the ocean, rivers and lakes, forests and fisheries, air and soil, sunlight and wind, coal and petroleum, animals and climate.

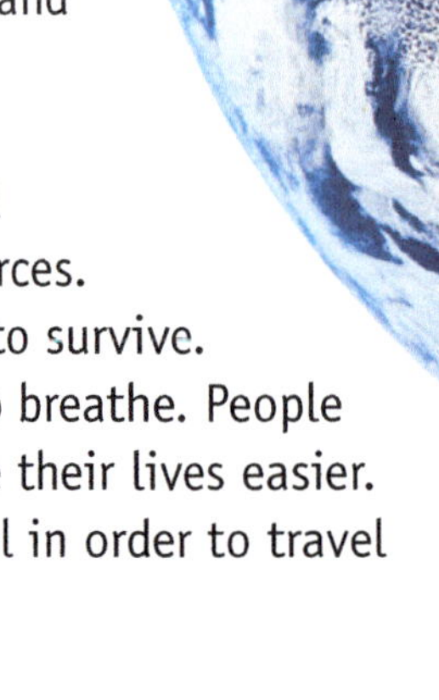

Those 'things' in your natural environment are called resources. People need some resources to survive. For example, they need air to breathe. People want some resources to make their lives easier. For example, they want petrol in order to travel in cars.

ISBN: 978-0170210348

The BIG issue

Sustainability (often referred to as sustainable development) is about using resources for your own needs and wants and making sure your future family will have enough of the resources for their needs and wants.

Sustainability can be about other things besides the environment. For example, you can have sustainable relationships. This means the relationships hold up. They don't break down into divorce between people, or wars between countries. However, when we use the word 'sustainability' today, it is usually to do with making sure Earth has a future.

What sustainability IS	What sustainability IS NOT
• A good life for everyone.	• A good life for just a lucky few.
• A good life now.	• A hard life now while saving for the future.
• Getting the economy going.	• Not worrying about the economy.
• Using the world's resources.	• Using none or all of the world's resources.
• Feeling OK about using resources.	• Feeling guilty about using resources.
• Protecting the environment.	• Treating the environment without respect.

Case Study The Brundtland Report

Gro Harlem Brundtland was the first female Prime Minister of Norway. In 1983 the United Nations asked her to head a commission to check out the global environment, and how to get sustainable development. The commission's report was called *Our Common Future*. It became known as *The Brundtland Report*. In it was an explanation of what sustainable development is. This explanation is the most quoted explanation used around the world today.

"Sustainable development is development that meets the needs of the present without compromising the ability of future generations to meet their own needs."

The report is called a milestone because it was an important event. After it came more talks about the environment and sustainability – meetings and agreements such as the 1992 Earth Summit, Agenda 21, the Rio Declaration, the Commission on Sustainable Development. They talked about issues such as the production of poisonous waste, finding new sources of energy, trying to get public transport in order to reduce pollution, disappearing water supplies, and using the imagination and courage of the world's youth to help get sustainable development.

ISBN: 978-0170210348

ISBN: 978-0170210348

1 Find the words or terms that mean the following:
- to hold up
- a supply you can draw on when you need to
- the whole world
- a big group of countries in an organisation working for world peace and a good future for all
- a group set up to talk about and find out about something
- weakening
- an important question that people can't agree on but look for a solution
- natural environment getting fouled by materials such as smog (air pollution caused by a mix of smoke and fog) and poisonous chemicals
- a programme run by the United Nations to get sustainable development in the 21st century.

2 Explain the difference between NEED and WANT. Give examples.

3 Use the following blocks to draw two walls. The first wall is a sustainable wall because it is made up of sustainable blocks. The second wall has fallen down because it is made up of non-sustainable (unsustainable) blocks.

taking more than your share	leaving the world better than you found it	water clean enough for people to drink	not thinking about future generations
using resources until they run out	being able to last	not damaging the environment	not being able to be continued
thinking about the future of resources	not putting the environment in danger	being careless and not protecting resources	helping future people to be able to meet their needs
air clean enough for people to breathe	holding up so it does not break down or disappear	looking after the planet	letting the last of a type of plant or animal die

4 In your group talk about which of the jigsaw pictures show actions that seem most about sustainability. Share ideas with the class.

5 As a class, come to an agreement about a meaning of sustainability.

6 Run a class competition for the best poster that uses the class meaning of sustainability.

ISBN: 978-0170210348

Greenhouse Gases - Sustainability's Enemies

Sustainability is closely linked to a

- set of gases known as greenhouse gases
- process known as global warming, or climate change.

gas = matter that is not solid or liquid, and which can keep expanding

process = action or series of actions

The Sun sends sunlight, and therefore heat, to Earth. Some sunlight is bounced back into space.

Some heat is released into space.

Some heat is naturally kept in the atmosphere by gases in the atmosphere, like water vapour.

Gases gather in the atmosphere. They wrap themselves around Earth like a blanket. The blanket traps the Sun's heat. In this way the gases act like a greenhouse by raising the temperature of Earth. They are called greenhouse gases.

Some human actions put greenhouse gases into the atmosphere. When this happens, the actions are said to emit greenhouse gases. Emit means to send out or to expel. A country or person who emits is an emitter. What is emitted is called an emission. The more greenhouse gases humans emit, the thicker the gas blanket.

This is called the greenhouse effect. If there was no greenhouse effect, people would not be alive. Earth would be too cold for people to survive. However, too much greenhouse effect can be bad for people, plants and animals.

As the blanket gets thicker, the Earth's climate heats up faster. This heating is called global warming and it is closely linked to climate change. Most scientists say climate change is bringing more storms and cyclones, more droughts and bushfires, more floods and landslides, an increase in temperatures, and a rise in sea levels.

ISBN: 978-0170210348

Greenhouse gases

- Some greenhouse gases come only from human actions. Examples are hydrofluorocarbons (HFCs), perfluorocarbons (PFCs), and sulfur hexafluoride (SF6).
- Some greenhouse gases are there naturally. Examples are water vapour, carbon dioxide, methane, nitrous oxide, and ozone.
- Some of these naturally present greenhouse gases are added to by human actions. For example, burning wood or coal puts carbon dioxide into the atmosphere. Flatulent (farting and belching) cows put methane into the atmosphere.

Greenhouse gases in the atmosphere

water vapour = water in its invisible gas stage (not solid or liquid).

carbon dioxide = carbon + oxygen, heavy, no smell, no colour, non-toxic (not poisonous), animals breathe it out, plants take it in.

methane = carbon + hydrogen, no colour, no smell, flammable (burns easily), large part of natural gas, non-toxic BUT displacing oxygen from air causes suffocation.

ozone = three oxygen atoms, pale blue, sharp smell, toxic.

nitrous oxide = laughing gas or happy gas, used as anaesthetic in hospitals, no colour, non-flammable, sweet smell and taste.

chlorofluorocarbons = carbon + chlorine + fluorine, non-flammable, non-toxic, man-made.

hydrofluorocarbons = hydrogen + fluorine + carbon, used in refrigeration and airconditioning, man-made, generally colourless and no smell with low toxicity.

perfluorocarbons = carbon + fluorine, non-flammable, non-toxic, colourless and no smell.

sulfur hexofluoride = fluorine + sulfur, colourless, no smell, non-toxic, non-flammable, used in electrical industry and for insulated glazed window filling.

ISBN: 978-0170210348

Case Study Governments and Greenhouse Gases

Governments of many countries are worried. They want to reduce the amount of greenhouse gases in the atmosphere. To look for ways to do that, they send representatives to special meetings.

An example is a meeting held in 1997 in a place called Kyoto, Japan. The representatives wrote a Kyoto Protocol. The Protocol said countries should reduce their greenhouse gas emissions. Before it could come into force at least 55 countries had to ratify it. Those 55 countries had to include all countries responsible for at least 55 percent of the industrial world's 1990 carbon dioxide emissions. At that time the world's top 10 emitters were China, US, European Union, Indonesia, India, Russia, Brazil, Japan, Canada, Mexico. It came into force in 2005. Only countries that ratify it are bound by it. The US has refused to ratify it. New Zealand ratified it in 2002. Like all the other countries in the Kyoto Protocol, its emissions have to be monitored and records kept.

A follow-up to the Kyoto Protocol was in Copenhagen, Denmark in 2009. 2500 climate experts from 80 countries put out a statement. "There is now no excuse for failing to act on global warming," it said. "Unless countries reduced their greenhouse gases emissions they would find it really hard to cope with climate change."

At the 2010 United Nations Climate Change Conference in Cancun, Mexico, governments showed they were determined to get to a low emission future together.

The New Zealand Government has many agencies carrying out work to do with climate change.
Every year it makes a greenhouse gas report for the United Nations.

An organisation that New Zealand started with 20 other countries is called the Global Research Alliance. It aims to make agriculture more sustainable. This means finding ways to grow more food without growing greenhouse gas emissions. A recent idea is to add fish oils to the feed of cows to reduce the amount of methane they produce.

ISBN: 978-0170210348

protocol = treaty

ratify = sign up to do as it says

US = United States of America

agriculture = producing foods and goods by farming

European Union (EU) = large economic and political group of countries in Europe

1. Explain the difference between man-made and natural greenhouse gases and then explain the meaning of the title of this unit.
2. List all the names of the greenhouse gases mentioned in this unit. Say whether each one is natural or man-made.
3. Make up a cartoon strip about two flatulent cows and greenhouse gases.
4. Draw a star diagram to show possible results of global warming.
5. You are the leader of a poor country that is losing land to rising sea level. Make notes about some things you will talk to the US President about when you visit Washington.
6. Find another five facts about the Kyoto Protocol.

ISBN: 978-0170210348

The Argument About Global Warming

The Issue	One Side	The Other Side
Is global warming for real?	Some people say yes. They are known as believers.	Some people say no. They are known as deniers – people who deny, and sceptics – people who doubt.
What is causing global warming?	Some people say it's a natural and normal change in climate.	Some people say it's human activities that are putting extra greenhouse gases into the atmosphere.

The BIG question

Should humans wait for scientists to prove whether it's nature or humans causing global warming, or should humans act now?

Thousands and thousands of scientists around the world are working on climate change research while you read this.

Scientists say some carbon dioxide comes from natural activity such as volcanic eruptions and people breathing. Earth can easily absorb that. The problem is the extra carbon dioxide from human activities like burning fuels such as coal. Earth can absorb around 97 percent of the methane released into the air. The problem is the remaining three percent.

Most scientists also agree that at the moment there are probably more questions about global warming than answers.

Most scientists agree that in the last 30 to 40 years, Earth has got warmer. They make charts to show rising temperatures over many years, loss of sea ice, the rise of sea levels, the increase of carbon dioxide in the atmosphere.

Climate research in Antarctica.

Scientists are humans. So they sometimes make mistakes. This gives ammunition to deniers and sceptics. An example is a mistake in the 2007 report of the Intergovernmental Panel on Climate Change (IPCC) which included false claims that Himalayan glaciers could melt away by 2035. In 2010 it admitted that the claim had no proof. It was based on a media interview with a scientist done years earlier. Scientists said the mistake did not damage the huge amount of evidence that showed the climate was warming and that human activity was largely to blame. It was one mistake in a 3,000-page report.

ISBN: 978-0170210348

Case Study: Global Warming and Archeology

Archaeologists study ancient peoples by examining the artefacts (man-made objects) they left behind. Global warming is causing some melting of ice sheets and glaciers. This is exposing ancient artefacts such as axes, bows and arrows used by ancestors of the Vikings to hunt reindeer. Much of the material of the artefacts, such as wool and leather, crumbles to dust in a few days unless it is stored in a freezer. Archaeologists are having to race around to find and save these precious artefacts.

Case Study: The Dung Beetle

Dung beetles love animal poo (faeces). They eat it and make baby dung beetles in it. They could soon become the latest weapon on Kiwi farms against global warming.

Scientists at New Zealand's Landcare Research have been studying the beetles. Recently they got money from the Ministry of Agriculture and Forestry's Sustainable Farming Fund to bring dung beetles into New Zealand.

Scientists say dung beetles will reduce emissions of the greenhouse gas nitrous oxide that livestock poo emits.

ISBN: 978-0170210348

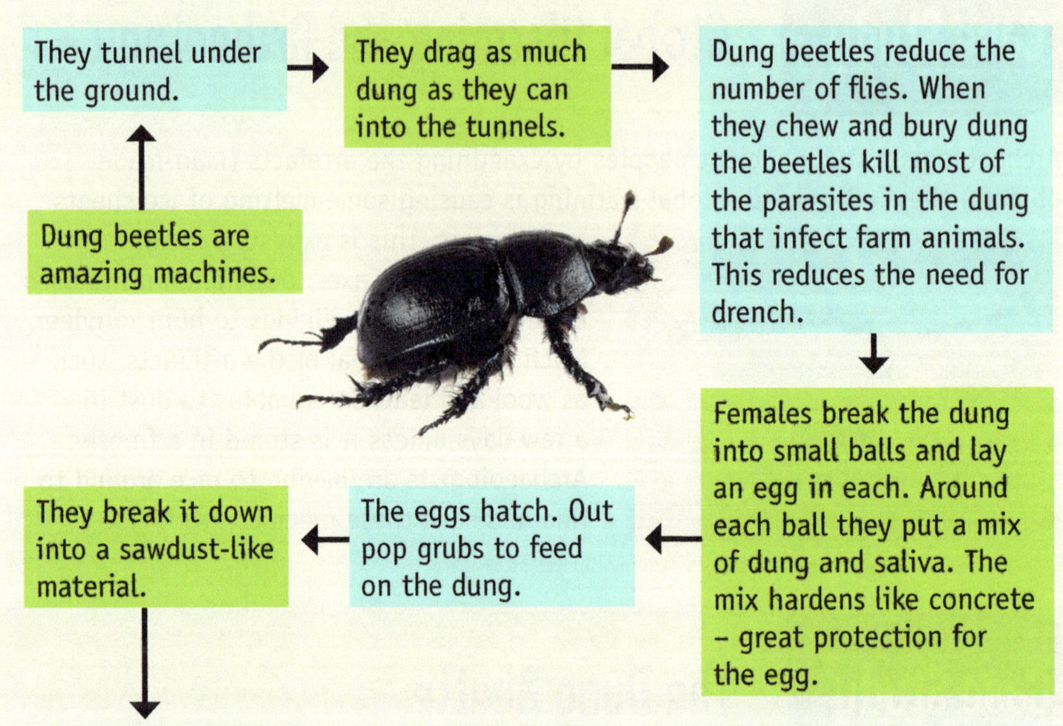

1 Make up five questions you would ask the world's top climate scientist if you were interviewing him or her for a newspaper story.

2 Make an advertisement for a farm magazine about a dung beetle for sale.

3 You are an archaeologist in Norway. An interviewer has just asked you about global warming. Give your answer.

4 Learn how to spell faeces (also feces), artefact (also artifact), leaching, sceptics, research, emissions.

5 Collect six opinions about the BIG question. Include your own.

ISBN: 978-0170210348

6 Explain what each of these pictures has to do with global warming.

ISBN: 978-0170210348

The Carbon Footprint

A carbon footprint is the emission of greenhouse gases caused by something or someone. It is often the amount of carbon dioxide emitted, although carbon dioxide is not the only greenhouse gas. A person, a household, a company, a country can have a carbon footprint. You can find your carbon footprint by using one of the free calculators online. It will ask you questions about your travel such as a plane trip or whether you bike to school or go by car or public transport, the food you eat, the things you buy, your recycling habits, the activities you do for fun.

There are different and complicated ways to measure carbon footprints. This is why you will find different figures given for the same country. Norwegian scientists, however, have recently created a single model to compare the carbon footprints of different countries. It allocates the carbon from the country of production PLUS the carbon from the country of consumption. This means it is possible to calculate the carbon cost of imported goods. This model gave the top three and the bottom three for a recent year.

Annual (yearly) emissions per capita (per person)		Estimation in tonnes
US	top 3	29
Australia		21
Canada		20
Bangladesh	bottom 3	1
Mozambique		1
Malawi		0.7

Whatever measurement is used, New Zealand is never in the top 10 but is always above average.

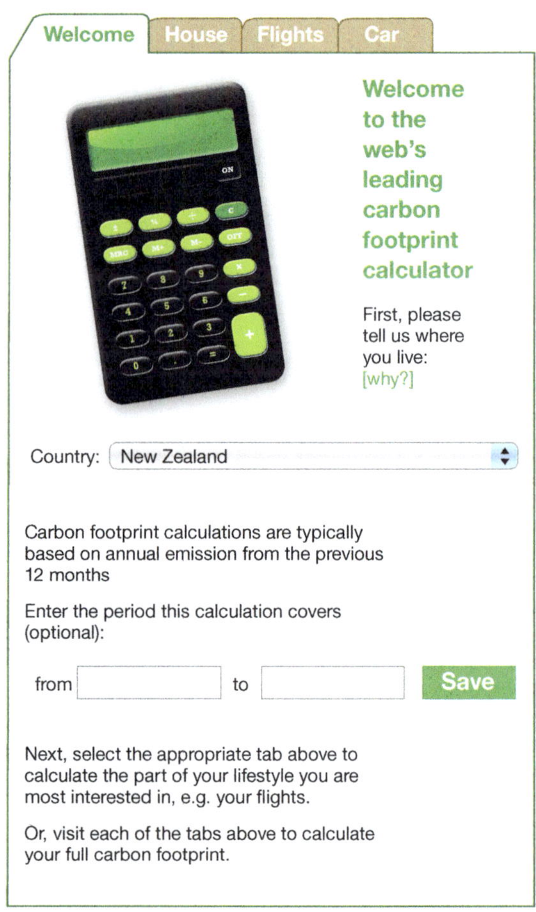

ISBN: 978-0170210348

Case Study 350.org

- 350.org is an international environmental organisation.
- It wants people all around the world to know about climate change.
- It wants the world to cut down on the number of tonnes of carbon dioxide it emits so that the concentration of it in the atmosphere is 350ppm (parts per million of carbon dioxide in the atmosphere). This is what scientists say is the safe limit. The ppm figure has stayed higher than 350 since early 1988. Each year it has increased. By 2010 it was over 390.
- 350.org is about measuring and data. The Observatory in Hawaii does the measuring. The National Oceanic and Atmospheric Administration (NOAA) of the United States publishes the measurements. It also puts out data on temperature. For example, it showed July 2010 was the second warmest July on record since 1880.
- It held the first 350 International Day of Climate Action on October 24 2009. Media described it as the most widespread day of political action in the planet's history. People held banners with 350 written on them, or formed 350 signs out of bodies standing together on the top of mountains, under the ocean, on streets, in deserts and snow.
- It made a big difference a month later in Copenhagen. By the time talks finished in Copenhagen, 112 countries had okayed the 350 target.

ISBN: 978-0170210348

Case Study The Stern Review

A leading worldwide expert on global warming and climate change is Lord Nicholas Stern. He is a professor at the London School of Economics. The UK Government asked him to review global warming and climate change. He wrote a 700-page document about it called *The Stern Review*. Now he takes his message around the world. He came to New Zealand in 2010. Here are some of the things he says.

- The benefits of strong, early action on climate change outweigh the cost.
- Climate change science, gathered over 200 years, is strong.
- For the planet to be sustainable, worldwide greenhouse gas emissions in the atmosphere must go from the current 50 gigatonnes to 35 gigatonnes by 2030 and to 20 gigatonnes by 2050.
- Rich countries should have to reduce emissions more than poorer (developing) countries.
- The world should develop new technology to reduce emissions.
- The world should fight forests being chopped down as forests are enemies of global warming.
- Much of New Zealand's emissions are from agriculture. New Zealand could lead the world in making changes in that area.
- Most countries could say they're not really responsible for worldwide emissions because they're only 1 or 2 percent of worldwide emissions, or even less. But countries have to pull together. Poorer countries could look at richer countries like New Zealand, and say, if those countries aren't bothering to reduce emissions, or finding it too hard to reduce emissions, what chance have we got? And why should we bother? So one country like New Zealand has a big effect on others.
- Countries and people should have to pay for their emissions.

ISBN: 978-0170210348

1. Draw a full page footprint. Label it 'The Carbon Footprint'. Fill the footprint with facts about carbon footprints.

2. In your group list ways for people to reduce their personal carbon footprints. Starters = use reusable shopping bags, hang washing out to dry rather than tumble dry, carpool, ride a bike, don't buy bottled water if your tap water is safe to drink.

3. Nicholas Stern uses the example of having a meal in a restaurant. We expect to pay for the space, for the service, for the food, for the energy, for the cooking and for the equipment. Yet we do not pay for the emissions associated with that whole process. Should we, do you think? Prepare a tweet or Facebook entry about this issue.

4. Imagine you are a member of 350.org. You are to give a short talk about what it is and why you belong to it at a school assembly. Make some notes for your talk.

5. The date 10/10/10 was a global climate working bee for everyone in the world to send a message to world leaders that people were ready for action on climate change. Find out about some of the events that took place then. What big events about global warming have taken place more recently?

6. Can you find out your carbon footprint?

ISBN: 978-0170210348

Carbon Offsetting

Maybe you can't or don't want to reduce your own carbon dioxide emissions. One way to get round that is to fund a carbon dioxide reduction somewhere else. For example, you could invest in a tree-planting scheme because planting trees reduces carbon dioxide emissions. Or you could invest in a wind power scheme because wind power does not pollute like burning coal pollutes. These actions are called carbon offsetting.

For example, Air New Zealand says it wants to help customers reduce the impact of their flights on the environment. It has a calculator for you to estimate the carbon dioxide emissions for your flight. You can offset the cost of your carbon emissions by making a donation to an Air New Zealand's Environment Trust project. Its first project is Mangarara Station in Hawkes Bay. The owners of the station want to develop a model sustainable farming system. Air New Zealand has given money to help plant 85,000 native trees and shrubs. You will be able to go and see the tree or trees your offsetting has planted.

An Emissions Trading Scheme is another version of offsetting.

ISBN: 978-0170210348

The New Zealand Government introduced its Emissions Trading Scheme (ETS) in 2008. New Zealand's ETS is linked to its responsibility to reduce emissions under the Kyoto Protocol. The government said it wanted New Zealand to be the world's first sustainable country.

The New Zealand unit (NZU) is the main unit of trade in the ETS. It will pay for one or two (depending on the time) tonnes of greenhouse emissions.

NZUs can be bought from the Government for NZ$25 each. Groups can also buy them from other groups in the scheme or from traders.

The diagram shows ETS at work using three industries as examples.

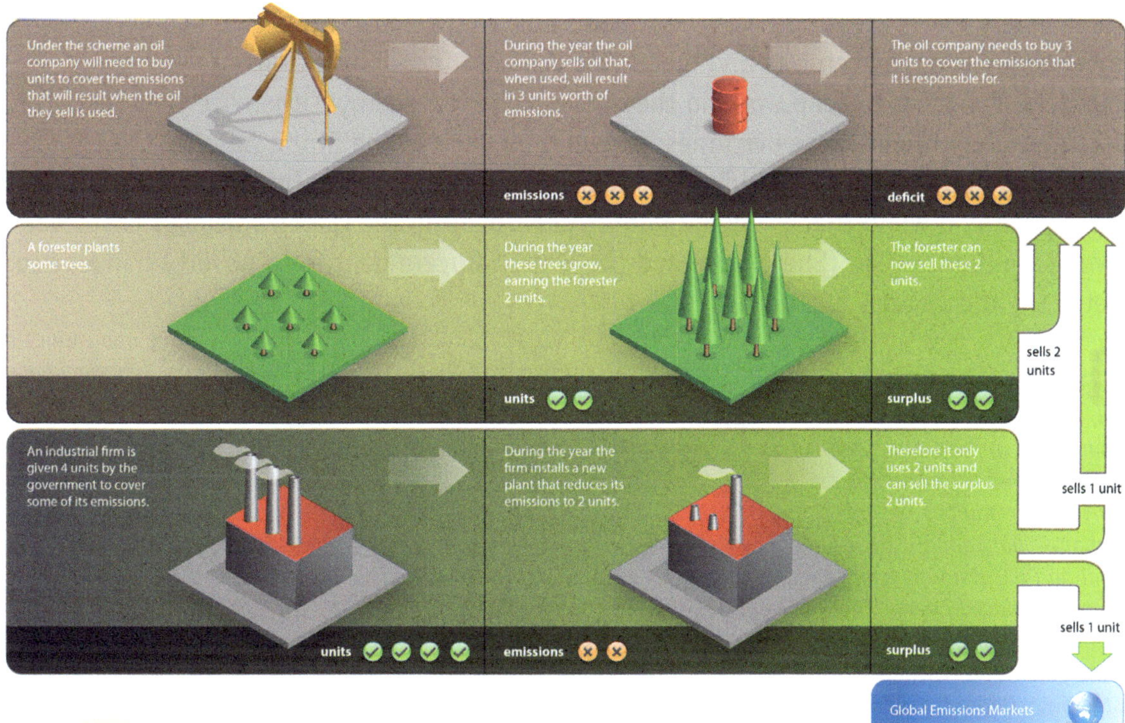

1. Find proof in the unit that 'offsetting' means to make up for, or to compensate.
2. Explain what a 'responsibility' is. Give an example from this unit.
3. Air New Zealand wants to be a leader of sustainability in the aviation industry. Suggest other ways it could do this.
4. Make your own copy of the diagram above. Under it, put all the terms to do with ETS, such as trade and unit, and put a meaning beside them.
5. 'One tonne' probably doesn't mean much to many people. It's hard to picture in your mind. In your group try to find a way of explaining exactly what it is.
6. Find out what is happening at the moment with the ETS.

ISBN: 978-0170210348

Sustainable Energy

Humans are clever creatures. They've worked out ways to save their own energy by inventing machines, vehicles and electronic devices. These inventions are energy-guzzlers. For them to keep working they have to be fed energy. Otherwise they just stop. Imagine life without these machines; it would be back to the Stone Age.

Any talk about sustainability has to include talk about energy. This is because humans have created a world that is energy-hungry. Just heating the Queen's palaces costs well over $2 million a year. (She once asked the British Government for a poverty handout to help. She did not get it.)

Fossil fuels are coal, natural gas, oil, peat. These energy sources come from ancient plant and animal remains (fossils) in the Earth. Over millions of years these remains changed into things like coal and oil. Humans found these things produced energy when burnt. The process goes like this. Burn fuel … heat water to make steam … steam turns turbines… turbines turn generators … you get electrical power. Fossil fuels also produce greenhouse gases that go into the atmosphere. And they are non-renewable. Once you've burnt a lump of coal, it is gone forever; another lump doesn't magically appear in its place.

The opposite of non-renewable energy is renewable energy. It's also known as alternative energy, green energy, or sustainable energy. What they mean is that the energy is not all used or exhausted. There will still be renewable energy for future generations to use.

Hydropower generates electricity from moving water. It produces little or no greenhouse gas emissions and the water is free. It is renewable.

Geothermal energy is renewable. It comes from the heat of the Earth. New Zealand has over 1200 identified geothermal fields. Its geothermal features are world famous. The Taupo Volcanic Zone and Yellowstone National Park in the US have the top concentrations of geothermal activity on the planet.

ISBN: 978-0170210348

Solar energy comes from the Sun. Photovoltaic (PV) cells in a panel are used to change it into electricity. As with all technology, you can still have problems with alternative energies. Recently, a badly designed luxury hotel in Las Vegas was covered in reflective glass. It condensed the rays of the sun and beamed them down on the swimming pool. Hotel workers called this the death ray. It melted plastic in the area and scorched the hair of a hotel guest.

Wave energy is produced when the wave rises into a chamber. It forces the air out of the chamber. The moving air spins a turbine which can turn a generator. In 2010 the New Zealand Government awarded funding to a project to harness wave energy on the Chatham Islands. The Islands were trying to reduce the use of diesel generation for electricity.

Tidal energy changes the energy of tides into electricity. When tides come into shore they are trapped in reservoirs behind dams. When the tide drops, the water behind the dam can be let out like in a river hydroelectric power plant.

Ocean Thermal Energy is still developmental. It produces electricity from the ocean by using the temperature difference between warm surface water and cold deep water.

Wind turbines use blades to collect the wind's energy. The wind flows over the blades creating lift, which causes them to turn. The blades are connected to a drive shaft that turns an electric generator to produce electricity. Experts say New Zealand has one of the best wind resources of any country in the world thanks to its location. It gets winds travelling across the ocean uninterrupted by other land.

Biogas is produced by the breakdown of organic (from plants and animals) waste in the absence of oxygen. For example, from manure on pig farms, and from sewage (liquid and solid waste from places like bathrooms and sinks). It is a mix of mainly methane and carbon dioxide. The gas can be used on site to produce electricity.

ISBN: 978-0170210348

Biomass is biological material such as waste and wood from living or recently dead things. It contains stored energy from the Sun. Energy from biomass can be converted into fuel. Biodiesel is made from vegetable oils and tallow (animal fats). Bioethanol is made from fermented sugars and starches, such as maize.

Nuclear power is said to be low-cost and low-emission. Turbine generators are driven by steam from water superheated by nuclear fission, when one atom splits into two. Nuclear power is said to be sustainable. However, there are worries about radioactive waste disposal, uranium, and the risk of accidents.

Case Study Offshore Wind Farm Issue

Wind turbines often get the NIMBY reaction – Not In My Backyard. Okay, say the scientists. Put the turbines out of sight in the ocean, or on big lakes. The wind often blows more fiercely and steadily there anyway, and maybe we can build them further and further offshore, perhaps even on floating platforms. Denmark is a world leader in wind farms. Its state-owned wind farm company has said it will stop putting wind farms onshore and put future ones offshore.

An offshore wind farm near the coast of Denmark.

ISBN: 978-0170210348

What critics say	What developers of offshore windfarms say
It will be hard getting the electricity back to shore.	The electricity generated simply travels back in electric cables. This means the turbines can be sited near coastal cities.
The noise from wind turbines will travel underwater and might disturb sea life.	Studies on the impact of noise from existing offshore turbines show the noise is very low frequency. Many species can't even hear it.
It will take too long to get the wind turbines up and running.	Building them, and later shutting them down, will take only about six months.
The turbines could cause coastal erosion.	The Danish experience shows no evidence of this. Scientists already recognise that global warming is eating at coasts through higher sea levels and more storms.
Storms could make turbines dangerous.	Offshore turbines are tested to stand up to storms. In stormy weather, the blades turn out of the wind and slow down.
They are ugly.	Most offshore wind farms are barely visible from shore.
They will disturb fishing.	They don't take up much space. They will provide safety for fish spawning and a place to hide from trawlers.
They will be a danger to ships.	Developers put wind turbines outside of established shipping lanes. If a ship goes off course, its radar will show the wind turbines. Wind turbines have warning devices to alert ships in bad weather.
Getting them built at sea will be a nightmare.	The transport of large wind turbine pieces such as tower sections and blades is much easier over water than on land. Ships handle large cargo better than trucks or trains. And there are no traffic jams.
They will kill birds.	Studies show the kill-rate by turbines is very low in comparison to other killing machines such as vehicles on roads, cats on the prowl, and glass in windows and doors.
They can't be put in deep water.	They can be placed in water up to 40 metres deep, and floating wind turbines can float in water up to 700 metres deep.
Wind is unpredictable and uncontrollable.	Yes, it is. For example, much of the electricity generated by wind in Denmark can't be used in Denmark. This is because, except for hydropower, electricity cannot be stored in large amounts. Power companies have to generate it when it is needed. But it might blow in the middle of the night when it's not needed so much. It might blow at the wrong speed to generate electricity. Yet it can still be used. Denmark exports most of its wind electricity through cables to Germany.

ISBN: 978-0170210348

1 a What advantage does geothermal energy have over solar energy?
  b Why is tidal energy more predictable than wind energy?
  c In what category of fuels would algae fuel belong?
  d How can email spam, which uses a lot of electricity, be reduced or stopped?
  e How did fossil fuel get its name?

2 Create a logo for non-renewable energy, and a logo for renewable energy.

3 Read the following and answer the question.

Biomass briquettes for fuel are blocks of flammable matter made from agricultural waste. Charcoal for fuel is made by getting wood to smoulder in buried mounds. In the Democratic Republic of Congo, cutting down forest for charcoal production is the biggest threat to the homes of mountain gorillas. Staff of Virunga National Park train people to make biomass briquettes. This replaces charcoal made illegally inside the park.

Use this information to describe how the sustainable energy issue has problems and solutions.

4 If you were Energy Minister for New Zealand, would you support offshore wind farms? Give reasons for your answer.

5 Make a summary of all the renewable energies mentioned in the unit. Find out which ones are used in New Zealand.

6 Choose one of the energy technologies from pages 22–24 to research.

ISBN: 978-0170210348

Sustainable Transport

How about this for a trip from hell – a monster nine day traffic jam stretching for 100km and in which drivers passed the time playing cards, sleeping on the asphalt, bargaining with food-sellers who came to make a killing, tramping off to find a place to toilet, worrying about fruit and vegetables in unrefrigerated trucks. It happened in the 2010 summer in China. On average in the months before the jam, vehicles on Beijing's roads multiplied by 1900 a day.

THE TRANSPORT PROBLEM

= 20-25 percent of world energy consumption
= 20-25 percent of carbon dioxide emissions
+ energy involved in making vehicles
+ pollution involved in making vehicles.

THE TRANSPORT SOLUTION

= sustainable transport, aka green transport
= sustainable energy use
= as near to zero pollution as possible
= low impact on the environment.

THE GOOD NEWS

Scientists and inventors are coming up with new ideas all the time.

Bikes are already sustainable but now they're getting even more streamlined. Recent designs include a bike that generates electricity, a folding bike that lets you carry it into lifts and up narrow stairs, another folding bike that even has wheels that fold, a bike that folds down to fit in a backpack, and so on.

ISBN: 978-0170210348

China's 3D Bus will run on tracks and be powered by electricity and solar energy. Its upper level will carry its 1400 passengers. Cars can drive under it. They will be warned when the bus is about to turn. The bus will have an inflatable escape ladder for emergencies.

One car idea is a car on a stick. It is a transparent solar-powered pod or bubble that can take up to four people and their shopping. The car has no driver and navigates by voice command, and satellite GPS navigation. When the car is not in use, a telescopic pole under it lifts it up to become a street light.

EN-Vs (Electric Networked Vehicles) were the big hit of the Shanghai World Expo in 2010. People can sit in them and be driven automatically. Using their cellphones, people can get the cars to drive themselves from where they are parked to where the people are waiting. These bubble cars balance on two wheels.

In 2010 Andre Borshberg of Switzerland made the world's first solar-powered night flight. Until then, most solar-powered planes had been unmanned and none had tried a night flight. Andre stored solar energy during the day and used it after nightfall. He said, 'We are on the verge of perpetual flight.' That would mean no more layovers in cities and the reduction or end of jet fuel.

ISBN: 978-0170210348

Case Study Solar Energy Race

Solar cars use photovoltaic cells (PVCs) to convert the Sun's rays into electricity. People can charge plug in electric vehicles through solar panels on their home, or on the roof of the car. Some electricity is stored in a battery so the car will go at night, on cloudy days, and in dark places.

The Global Green Challenge encourages research into solar energy. Entrants design and build a car to go through the Australian outback with only sunlight as fuel. Every two years teams from around the world set off from Darwin to Adelaide. They drive between 0800 and 1700 hours and then set up camp along the roadside. Some teams have high-school students in them. Many go on to careers in sustainable technology. The 2009 winner was a Japanese team from Tokai University. Using compound solar cells, they travelled at an average speed of 100km per hour. That year was the first year of another race section – one for electric, hybrid and other alternative energy low-emission vehicles.

Case Study Monorail

A New Zealand company called Shweeb Holdings won a US$1million award from internet giant Google in 2010. It beat over 150,000 ideas and inventions over a two-year search around the world. Google was to get a shareholding in the company and put its profits into a charitable trust for improving public transport.

ISBN: 978-0170210348

The award money was to help the company develop its monorail which it had run at a Rotorua adventure park for several years. The system will be used in cities around the world.

The system has transparent pods travelling on overhead cycle monorails. Users pedal their pods like bikes. They lie down to do this so they use only half the energy they would need if they were riding a bike on the road. They travel at about 20km/h at about 6m off the ground. They can't crash into anything or fall from the rail. The ideas man is an Australian who likes biking and is interested in sustainable transport.

1. Explain what the underlined terms mean.

2. Run a class competition to find the best name for the car on a stick.

3. Describe how you could combine solar and monorail technologies.

4. The Delft University of Technology in the Netherlands has unveiled its latest green vehicle. The electric Superbus can carry 23 passengers. Its designers say the bus is a future vision of sustainable public transport. It has an aerodynamic shape made from lightweight carbon. It is powered by four electric motors that can get it to a top speed of 155 mph. Give evidence to show its designers wanted to combine high-speed with sustainable travel.

5. The European Environment Agency held workshops in Denmark to think about how society might get on in the future without air travel, or with much less of it. (Participants flew to attend.) Talk about this issue in groups and see which group can come up with the best ideas.

6. Find out about hybrid vehicles that combine two or more sources of power.

ISBN: 978-0170210348

Overpopulation

When you go to bed each night, Planet Earth has about 200,000 more humans to feed and house than it had the day before. This increase of humans is known as population growth.

Overpopulation is when the population is growing too big for the environment to cope with. There are not enough resources. Or there are too many people for resources to stay sustainable and not run out.

1900 = 1 billion people in the world

1960 = 3 billion people in the world

2000 = 6 billion people in the world

Best guesses, known as projections or projected population, for the future is that some time between 2040 to 2050, the population will be between nine to 11 billion.

Causes of overpopulation:

- more people being born
- fewer people dying
- resources running out
- resources being overused and not being kept sustainable.

Results of overpopulation:

- resources run out more quickly
- people argue over and fight for resources
- overcrowded cities
- people are not as comfortable as they used to be (known as lowered lifestyle)
- not enough fresh water
- air and water get dirtier
- forests get chopped down
- farmland gets polluted or turns into desert or is taken over for housing
- some types (known as species) of plants and animals become extinct (no more)
- more babies and young children die (known as high infant and child mortality)

ISBN: 978-0170210348

- more factory farming which can spread diseases in animals including humans
- new global diseases (known as epidemics and pandemics)
- starvation and poor diets
- more old diseases such as rickets (bones get soft) caused by poor diet
- low life expectancy (age at which people can expect to live to)
- unhygienic (dirty) living conditions
- more crimes such as people stealing resources to survive
- governments pass laws that take away some personal freedom
- people come up with interesting solutions such as putting humans on the moon.

Possible solutions:

- control the birth rate in countries where it is out of control
- make sure people know about and can get birth control
- educate people so they know the results of overpopulation
- make sure people in all countries can get to family-planning clinics
- pass tax laws that reward couples who have no more than say two children
- improve health-care so people no longer have to have extra children in case some die young.

Case Study India

Official estimates (best guesses by experts) say 77 percent of Indians live on less than US$2 a day. Nearly half the children do not have enough to eat. Yet India is expected to have 1.6 billion people by 2050. At that time China would have 1.4 billion and the US would have 439 million.

Big cities in India equals big problems. If you live in certain parts of the capital Delhi, it is likely you would have to get up at four o'clock in the morning to fill a bucket with brownish water from a tap that gives water for only half an hour a day. You would be always fighting with your neighbours to get water. Your clothes and the masks you wear as defence against the dirty air would get black very quickly. You would probably have a breathing disease such as asthma. You would see people living in slums and see more slums develop. Your life might end in a road accident as your city has a road accident death rate of six people a day – the highest in the world. And you could find many statistics to prove how unsustainable things are at the moment. How about this one? It is estimated that each cow in Delhi, and cows being sacred are still allowed to wander in streets, has an average of about 300 plastic bags in its stomach.

ISBN: 978-0170210348

Case Study China

The Policy

In 1978 the government of China set up a one-child policy. 'One is good,' it said. 'Two is okay, three is too many.'

Exceptions

Parents in the countryside, especially if their first child is a girl or is mentally or physically challenged, have sometimes been allowed to have another child. So have parents who have no brothers or sisters, parents who come home from abroad, and parents who lose their child in a natural disaster such as an earthquake. Wealthy couples might be able to pay a fee to the government to have more than one child.

Results

- Critics say it has led to the killing of female babies (as most parents want their only child to be a boy), abortions, gender imbalance where there are more males than females, children abandoned by parents, children taken away from parents by family-planning officials.
- People who disobey are fined. They may not get the same government benefits as parents with only one child.
- Family-planning officials work hard to explain to parents why China needs to follow the policy.
- Government says the policy has reduced population growth, although there are still a million more births than deaths every five weeks.

ISBN: 978-0170210348

1 Carefully look at these population figures.

Population	
China	1,339,190,000
India	1,184,639,000
US	309,975,000
Indonesia	234,181,400
Brazil	193,364,000
Pakistan	170,260,000
Bangladesh	164,425,000
Nigeria	158,259,000
Russia	141,927,297
Japan	127,380,000

- **a** These figures were for populations in 2010. How are they presented?
- **b** Can the figures be called projected populations? Explain your answer.
- **c** Put the figures on a graph.

2 Explain the links between resources and overpopulation and draw a diagram to show the results of overpopulation.

3 Issues are important topics that people have opinions about. An example is abortion. Another example is the rights and responsibilities of a person. As a person you have the right to make your own decisions. At the same time you have the responsibility to be a good citizen for your country. In your group make up a short role-play of a public meeting where a family-planner is explaining a government's one-child policy.

4 In 2010, New Zealand had the 121st highest population out of 192. Find out what experts say about New Zealand's population in relationship to sustainability.

ISBN: 978-0170210348

Environmental Refugees

Environmental refugees are climate refugees. They are forced to move from their homes because their environment is unsustainable. Natural disasters are not new. What is new, say experts on climate change, is the possibility of a big increase in the number of natural disasters. They could put tens of millions in the developing world (the poorer countries) on the move. These refugees leave with little more than what they wear. Their movement is fast and unplanned. It will put even more stress on places that are already unsustainable.

Case Study Hungary

In October 2010 a reservoir at an aluminium plant in Hungary burst its banks. A million cubic metres of red muck formed a 3m wave of poison. It raced over three villages and on to the Danube River. It killed at least four people and injured 150 others. If it had hit the villages at night, most people would have died. The muck, known as sludge, covered 41 sq km. It had a high content of heavy metals. Water in the Danube was immediately polluted and fish began to die. A tributary of the Danube was declared dead. Hundreds of Hungarians became environmental refugees.

ISBN: 978-0170210348

Case Study Bangladesh

Bangladesh has a population that is growing at an unsustainable rate. It is also one of the most densely populated countries of the world. That means more people on average live in a Bangladeshi square kilometre than people live in square kilometres in other countries.

About 80 percent of its people live on a floodplain and delta. Much of this land is less than a metre above sea level.

Climate change has brought more and worse storms and floods than usual, a rise in sea level, increased rainfall, increased melting of snow from neighbouring mountains, salt damage to crops, erosion to river banks. Up in Nepal and the Himalayas people have been clearing forests. This adds to Bangladesh's problems. It increases runoff, causes soil erosion, harms the ability of land to absorb water, brings more flooding.

The mosquito population is rising. Diseases like malaria and dengue fever are spreading.

When half of Bhola Island became permanently flooded, experts described the people who had to leave as some of the world's first environmental refugees.

Bangladesh's capital Dhaka is the fastest-growing megacity in the world. Yet this is where many environmental refugees head.

About half the people of Dhaka live in slums and shanty towns. Homes are made of plastic sheets, waste bamboo, corrugated iron. They are near rivers and markets, and filled with garbage and streams of raw sewage. Among the hordes scuttle cockroaches and child traffickers.

Wetlands and other water bodies disappear as Dhaka fills them in to build multi-storied buildings to house new arrivals.

Dhaka is known as the Rickshaw Capital of the World. Hundreds of thousands of them zip through streets. Cycle rickshaws are non-polluting. But they cause so much congestion they've been banned from much of the city. Auto rickshaws with their two-stroke engines pollute. The government is trying to replace these 'baby taxis' with 'green taxis' that run on compressed natural gas.

Dhaka's air pollution has been measured as being the highest in the world. Vehicles are old and poorly serviced. Their emissions mix with dust from roads and building sites and toxic fumes from industrial sites. The government is working to get cleaner technologies in the brick industry and to reduce air pollution.

People don't have basics such as electricity and clean drinking water.

ISBN: 978-0170210348

Dhaka, Bangladesh.

1 Draw labelled sketches to show the meanings of the following: megacity, decade, stress, densely populated, floodplain, delta, runoff, shanty town, capital, basics, tributary, reservoir.

2 Explain what a rickshaw is and the different types of rickshaws. Then give your opinions about why rickshaws are not a feature of New Zealand and whether or not they should be.

3 Look at the map and case study on page 36. Answer these questions.
   a From which direction would snow melt come?
   b Which river flows through Bangladesh as the Jamuna?
   c Name Bangladesh's two neighbours.
   d Name the largest bay in the world.
   e Name the ocean in which that bay is situated.
   f Name the flooded island.
   g From which direction would cyclones come?
   h Describe the elevation (height) of most of the land.

4 Make your own copy of the map. Put on it extra information. For example, put the label 'largest bay in the world' where it belongs.

5 The United Nations Environment Programme defines environmental refugees as 'those people who have been forced to leave their traditional habitat, temporarily or permanently, because of a marked environmental disruption (natural and/or triggered by people) that jeopardized their existence and/or seriously affected the quality of their life'. Put this definition into simpler language.

6 Find out why some people in Africa are becoming environmental refugees.

ISBN: 978-0170210348

Overconsumption

While people in parts of the world like Bangladesh struggle to survive, other places have people suffering from affluenza. Affluenza is a clever play on words. A word for wealth is affluence. The full word for what we call flu is influenza. Affluenza is the 'disease' you get when you keep buying stuff. Rather than making you happy, it can make you feel unhappy. Affluenza is another term for overconsumption.

Consumption is the opposite of production

You produce food (grow it in your garden).	You consume food (eat it).
You produce water (collect it in a rain barrel).	You consume water (drink it).
You produce money (earn it by doing an after-school job).	You consume money (spend it).
You produce electricity (put solar panels on your roof to get energy from the Sun).	You consume electricity (turn on your computer).
You produce goods (such as the hoodies your mother makes in a factory).	You consume goods (buy clothes).
You produce services (such as the plumbing your father does for a job).	You consume services (hire people like plumbers to do jobs for you.)

Overconsumption = using too much
= using resources at a faster rate than they can be replaced
= not a sustainable way to live
= an event mainly in wealthy countries.

One person in the United States generally consumes 30 times more than what one person in India does. Yet surveys have shown that Americans were most happy when they had way less stuff. No home computers, no cellphones, no video games, no emails, no internet. Today the United States is five percent of the world's population. It consumes over 30 percent of its resources. That is overconsumption.	The world's poorest 20 percent consume 1.5 percent. The world's middle 60 percent consume 21.9 percent. The world's richest 20 percent consume 76.6 percent.

ISBN: 978-0170210348

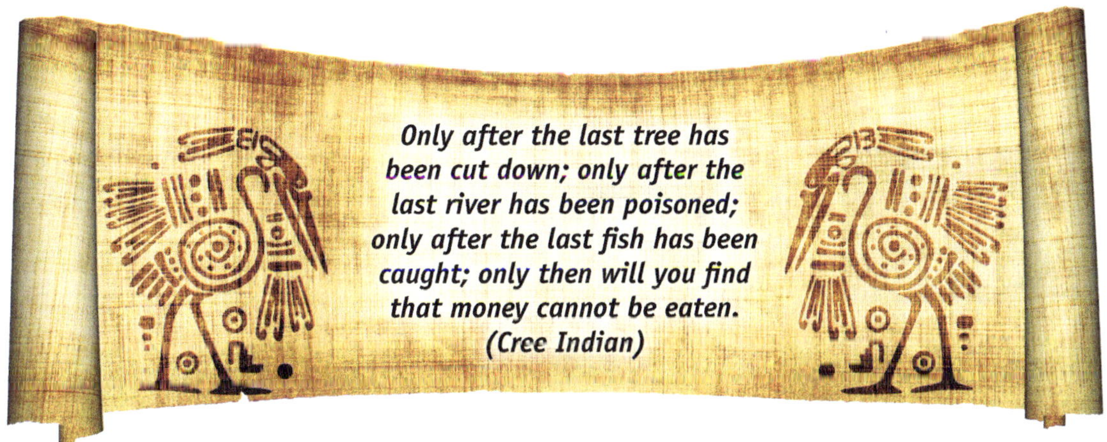

What helps turn you into an overconsumer?

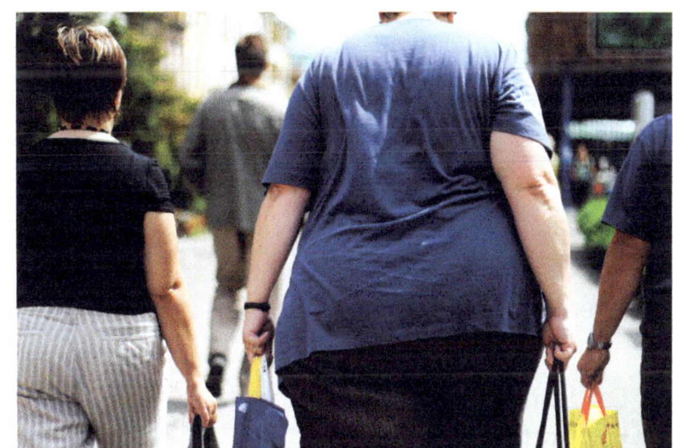

- There isn't an overconsumer gene. You get taught to be an overconsumer. Once countries moved into mass-production, people had to be taught to want stuff and keep on wanting more stuff. Otherwise nobody would buy all the stuff being produced.
- Goods are displayed so they catch your eye.
- Advertising and marketing is BIG business.
- Children in front of televisions make a captive audience for advertisers.
- Fashion is always changing.
- It is easy to buy goods with credit cards.
- The invention of shopping malls encouraged buying.
- Ideas change. Years ago people admired those who did not overconsume; today images of overconsumption are everywhere. Does a single movie star really need a house with 50 rooms?
- Experts in various fields often think in terms of stuff. For example, some experts on children encourage parents to give children their own stuff so the children learn pride of ownership.
- In the past, people did not throw stuff away. If their tools got damaged they fixed them. If their clothes got torn and worn they mended them. If they grew out of something they passed it on to someone else. When machines began mass-producing, producers came up with the idea of planned obsolescence. This means they planned their products not to last forever. The products soon became obsolete – old-fashioned and without all the gimmicks and gadgets the new versions had. So people were encouraged to throw them away and buy new stuff.

One suggestion to solve overconsumption is to find another planet for humans to live on. In 2010, scientists said they had found the best candidate to date, a planet known as Gliese 581g. It is about 193 trillion kilometres from Earth. It would take 20 years travelling at the speed of light to reach it or several thousands of years in a spacecraft built using the best rocket technology around.

ISBN: 978-0170210348

Case Study Food and Overconsumption

Around the world about 25,000 people die from hunger each day. At the same time many people move into the category of obesity – being really overweight.

Overconsumption of food is one cause of obesity.

This is not happening just in developed countries. Recent obesity figures show a high figure for developing countries such as Colombia and China.

Food producers spend millions of dollars advertising junk food and weird diets but not on educating people to eat less and eat better.

Research shows watching TV encourages overconsumption of food.

Medical resources are used to deal with the health costs and results of obesity such as liposuction.

Cattle-raising has led to the clearing of rainforests in parts of the Amazon. The meat produced is not to feed local people but for fast food restaurants such as McDonald's.

Supermarkets display food from all around the world. Want to eat summer fruits in the middle of winter? Go to your supermarket. The fruit may taste like cardboard, cost megabucks and have a monstrous carbon footprint but it will be there.

Researchers have found that obesity, like drug addiction, is linked to the brain's reward system. Overconsumption can trigger a gradual increase in the reward's lowest limit. Over time, the person needs more and more food to satisfy the craving. Researchers also found that increased or unlimited access to high-fat foods, can trigger overconsumption.

Healthier food with lower calories is more expensive than junkier food. This can be an issue for some families.

The food industry markets its products well. It makes food so tasty that you can hardly resist it. It prices products just right, and puts them everywhere. Once, people went to a petrol station and there was nothing to buy but petrol and oil. Today, if you were locked inside one you could survive for a long time because of all the food and drink there.

ISBN: 978-0170210348

1 List things you consume. Think of what has gone into making your stuff such as the trees to make the paper on which you are writing. Decide if you suffer from Affluenza. Be prepared to defend your answer.

2 One suggestion to stop obesity is to narrow the doors of junk food outlets so that obese people can't get in. What do you think of that suggestion? What are some causes and results of obesity?

3 What are some causes of overconsumption of products (excluding food)? Generally, is overconsumption a good thing or a bad thing? Give reasons for your answer.

4 Choose one of the following to be on a poster you make to warn about the dangers of overconsumption.

> You can't have everything. Where would you put it?
> *Steven Wright*

> Our world has enough for each person's need, but not for his greed.
> *Gandhi*

5 In your group, create a short play about planned obsolescence.

6 Make some notes to contribute to a discussion about whether the planet should come up with solutions to overconsumption or rely on finding another planet for humans to live on.

ISBN: 978-0170210348

Life Cycle of a Cellphone

Every product, whether it's as simple as a pencil or as complicated as a computer, goes through a process like this.

Design

What will the cellphone look like? What materials will be used to make it? What colour will it be? Will it be cheap so it doesn't last as long as a more expensive one? Will it have modular parts so that if one part breaks you can fix the broken part and not have to throw the whole phone away? Will it come with accessories such as a face-plate? Will it have only one application, or will it have 20 applications?

Materials extraction

This means extracting (getting out) the materials from where they are found. Often this involves mining. The figure usually quoted is that one cellphone needs about two kg of materials. That includes over 20 metals, such as cadmium, lead, nickel, mercury, manganese, lithium, zinc, arsenic, antimony. Some materials are toxic. Extracting them uses a lot of energy and can cause pollution and waste.

Materials processing

This means working on these materials to turn them into what cellphones need. For example, crude oil is combined with natural gas and chemicals in a factory to make plastic. The circuit board, liquid crystal display, and battery all have to be made in factories. More energy, more pollution, more waste.

Manufacturing

This is putting the processed materials together to make the cellphone.

Packaging and transport

Packaging to display the cellphone consumes resources such as paper from trees, plastic from crude oil, aluminium from ore. More energy, more pollution, more waste. A plane, truck, train or ship will take the cellphone to a shop near you. More energy, more pollution, more waste.

Disposal

This means getting rid of the cellphone once you've finished with it. Throwing it away puts it into a landfill or an incinerator. There, valuable raw materials such as gold and copper are lost. Toxins from the cellphone can seep into the environment and get into the food chain to damage plants and animals including people. Recycling the cellphone is the sustainable way to go.

ISBN: 978-0170210348

You can make your cellphone more sustainable.

- Keep it longer.
- Look after it so it lasts longer.
- Charge your battery properly.
- Take it back to the store when you've finished with it, for re-use or recycling.
- Use new technologies such as solar-powered batteries and hydrogen fuel cells.
- Unplug the charger so it does not become an energy vampire taking energy while plugged into a wall socket but not plugged into your phone.
- Use a charger that automatically senses when the phone is not plugged to the charger and cuts the power supply from the wall socket.

Some companies are producing more sustainable phones.

Built-in pedometer that works out savings on carbon emissions from walking versus driving in a car and then presents the data in the number of trees you've saved.

Body of plastic made from recycled water bottles.

Powering the battery through a hand crank.

In-phone electronic manual.

Completely biodegradable. Made largely of bio-plastic developed from natural resources like corn. You can actually bury the print board, battery, and antenna and they will disintegrate.

Seventeen of the world's largest cellphone companies signed an agreement that says a majority of new handset models were to include a universal charger by January 1, 2012. No more having to throw or give away your charger when your phone dies. It was estimated that could cut energy consumption by as much as 50 percent globally and could reduce greenhouse gases from between 13.6 to 21.8 million tonnes per year.

Motorola is paying to offset the carbon emissions from manufacturing, distributing, recycling, and using its new green phone.

Making commitments. Sony Ericsson, for example, is committed to reducing carbon dioxide emissions by 20 percent and greenhouse emissions by 15 percent from its cellphone's lifecycle by 2015. It joined a group working to get the European Parliament to ban the use of environmentally-unfriendly things in consumer electronics from 2015 onwards.

Screen light sensor to use less power.

No environmentally-unfriendly things like beryllium and brominated flame retardants that often appear in cellphones.

Solar-powered touchscreen.

Small and light packaging made from recycled paper.

Application to suggest how to lead a more environmentally-friendly lifestyle.

ISBN: 978-0170210348

1 Find the words that mean the following: burning waste, mechanically, made with separate units, taking out, a hand tool to start something, promising to do something, thing produced by effort, treated and refined, a programme that gives instructions, unrefined, devices using the movement of electrons, able to break down into compost, poisonous, leak through, making available, kilogram of one thousand grams.

2 A cellphone has hidden environmental impacts. True or false? Give evidence to support your answer.

3 Learn the six parts of every product's life cycle.

4 List the nine metals mentioned in materials extraction. Find out if they are toxic or not.

5 Less than one percent of mobile phones are recycled. With your group, plan a campaign to encourage students to recycle rather than throw away their old phones.

6 Choose any product that you use. Find out about the six parts of the process of its life cycle.

ISBN: 978-0170210348

The Poo Problem

In the 18th century (1700-1800) people in Britain began to invent complicated machines driven by steam-power. They put many of the machines into factories and used them to make things. This was called the Industrial Revolution. It spread to North America and Europe. Suddenly the environment had enemies such as pollution and industrial cities. The population of England's capital city, London, shot up to over a million people.

Imagine the amount of urine and faeces (pee and poo) produced by so many people. Rich people put theirs in chamber pots (bedpans) which servants carried off to empty in cesspits. Common people used outhouses or buckets. At night men drove horses and carts through the city. They bucketed the human waste that they called night soil out to their carts from backyards and cesspits, and carried it to farms outside the city. Farmers spread it on their farms. This stopped in 1870 when Britain began to get cheap guano (bird droppings) from South America for plant fertiliser. Meanwhile in the new cities men had started to dig open gutters to carry sewage.

After sewage is treated, in an effort to make it less polluting, liquid and solids are left. The liquid is called effluent. Solids are called sludge. You can find thousands of toxic stuff in sludge. Arsenic, lead, mercury, bacteria, viruses, parasitic worms, asbestos ... What to do with it? If you incinerate it, you put pollution into the air. If you put it in rubbish dumps, you get leaching (liquid pollution) into groundwater. If you put it in the ocean, you make huge dead seas.

Into these gutters went sewage. It was industrial waste and chemical waste mixed up with human waste. Untreated, it plopped out into rivers (How safe would water from there be to drink?) and oceans (How safe would that water be to swim in?). Countries have been spending vast sums of money ever since trying to solve the sewage (called black water) problem. When sewage systems become sustainable, we can call sewage a resource and not a problem.

ISBN: 978-0170210348

Case Study Cultural Differences

Many homes in India have flush toilets but millions don't. Millions have bucket toilets. Millions of people still do their business outside. Some schools have no running water and no toilets. A recent United Nations report showed more people in India have access to a cellphone than a flush toilet.

The River Ganges is India's holy river. Thousands of people bathe in it each day. Yet it is one of the most polluted rivers in the world. It's a receiver of human faeces and urine.

Some public toilets have as many as 400 seats. The females who clean them with brooms and tin plates are known as scavengers. They put the waste into baskets and carry them on their heads to a dumping place that could be up to four kilometres away. Baskets aren't water-proof. Imagine what happens to the waste on the human head in the rainy season. The government has passed laws to stop this scavenging work but it still goes on because females need the work to survive.

In some places villagers get a grant from the government if they build outhouses on their land. Some villagers build the outhouse, use it as a storeroom, get the grant and continue to use the ground outside as their toilet. In some places teachers don't get paid unless they have a plumbed-in toilet in their house and encourage their students' families to do same. But shortage of water means many people can't afford to waste water in a flushing toilet.

Dharavi is a slum of Mumbai. It featured in the 2008 award winning film *Slumdog Millionaire*. At that time there was only one toilet per 1,440 people there. Mahim Creek was widely used by locals for urination and defecation. The Indian government has plans for Dharavi to have sustainable development.

The United Nations report said it costs about $300 to build a toilet. That includes materials, labour, advice. The return is between $3 and $4 for every dollar spent, through less illness and more work.

ISBN: 978-0170210348

Case Study Solutions

Arriving in the world are sustainable sewage systems that produce water clean enough for non-drinking use and which can be safely put into the environment. One uses plants such as water hyacinths, reeds, rushes, lilies, and duckweeds. They break down toxins through their root microbes. They gather heavy metals in their stems and leaves. These plants provide an environment for bacteria, fungi, snails, and fish to live and do their bit to make clean water. This system is also known as an artificial wetland. Another solution is the solar aquatic system. The wastewater goes through a series of tanks in a greenhouse, where plants and animals break down the sewage.

Biogas plants are being developed to capture gas from sludge. Sewage is piped into a sealed container called a digester. There it produces a mixture of methane and carbon dioxide along with slurry (thin mixture of liquid and solid). The slurry is harmless and has no smell. It can be used directly on the land as a fertiliser. The methane is piped out the top of the digester.

A British water company has a Volkswagen Beetle powered by methane gas that comes from sewage treatment. It's called the Bio-Bug. The waste from 70 homes can power the Bio-Bug for a year.

A Swedish inventor has designed the peepoo. It's a biodegradable plastic bag. A person makes a deposit in the bag and buries it. Urea crystals coating the bag sterilise the waste and break it down.

ISBN: 978-0170210348

Case Study Freedom Campers in New Zealand

Many tourists come to New Zealand to enjoy its greenness and cleanness. Then they rent vans without toilets. Where do they toilet? Outside in public places.

Their actions make locals angry. Shoot them, say some. Not an option - New Zealand has human rights laws. Make them eat their own waste, say others. That would also be against human rights. Ban vans without toilets? Tourists would just rent cars and sleep in them. Create environment officers to police areas and deliver instant fines? The officers would need to find people in remote places, and then catch them in the act.

One solution is to educate people about the correct way to dispose of human waste.

People can use Poo pots. They come with biodegradable cornstarch bags. The bags can be dropped into a toilet, even a flush one. You then disinfect and reuse the pot. Or they could use a composting toilet that turns the waste into compost.

1. Explain the differences between chamber pot, cesspit and outhouse.
2. Work out what two actions the following two would be associated with: allowed to work only between midnight and five a.m., and females needing the work to survive. Building toilets in houses makes actions like those unnecessary. List other results of building toilets in houses.
3. Hinduism is India's main religion. Hindu text bans defecation near dwellings. Officials say it is hard to convince uneducated villagers to stop ancient customs and use an enclosed bathroom. In your group talk about possible ways to sort out clashes between tradition and modern ideas of environmental sustainability.
4. In the US sludge was renamed biosolids and sold to farmers as fertiliser. Only one to three percent was good for plants. The rest of it was polluted waste. If you knew this and were a seller of sludge, would you continue to sell it? Give reasons for your answer.
5. You rent out campervans. Make up the notice you will put on the inside of the door to encourage users to dispose of human waste safely.
6. Chlorine bleaching to get white toilet paper produces dioxins that are very toxic chemicals. Find out about them.

ISBN: 978-0170210348

Waste, Garbage, Junk, Trash, Rubbish, Refuse

Think of all the other things beside toilet deposits that become your waste – the pen that ran out, the computer that died, your lunch wrapper, the smashed iPod, your old toothbrush ... and you are just one person. People leave waste in the Himalayas and down in Antarctica, up in space and in the ocean. Big consumers produce big piles of waste. An American on average produces about twice as much waste as anyone else. Waste is the enemy of sustainability. This is why you always hear 'Reduce, reuse, recycle'.

Reduce means stop producing so much waste.

Reuse means using things again instead of throwing them out.

Recycle means putting things into a recycling bin so they can be broken down and materials in them reused.

For example, a computer has valuable resources in it such as copper and gold. It also has heavy metals such as lead and mercury. So you don't want your computer to end up in a landfill where it can damage the environment. You could recycle it by taking it to an eDay. The 'e' stands for recycling of electronic gear. If the computer still works, you could reuse it by giving it to someone. These actions mean you will reduce waste.

However, pick your recycling place. Guiyu in China is the biggest electronic waste site on the planet. It gets much of this from other countries such as the US. It's called the electronic graveyard not only because it gets dead e waste but because if you work there you could be hurrying towards an early grave. Workers use bare hands to strip out metals and chips; cook circuit boards, wires, and other plastics to get other metals; use dangerous acids to get gold from microchips. It's all so polluted you can't drink any water from there – water has to be trucked in from somewhere else.

The oldest way to get waste out of sight is to throw it into a rubbish dump, now called a landfill. A sustainable landfill would clean itself. It would make sure anything that caused greenhouse or toxic emissions stopped doing so within a generation. Nasties in

ISBN: 978-0170210348

the landfill would get locked up or broken down into harmless stuff. Each generation would inherit a clean environment. Not there yet, say the scientists, but working on it.

In the meantime, leachate continues to be a problem. That's the liquid that leaches – gradually goes somewhere, like into soil. When it rains or floods, water gets into landfills. It mixes with toxic substances that are leached from waste. The resulting leachate pollutes waterways. Not good for anyone or anything. Not good for New Zealand's clean and green image.

Landfill gas is a mix of methane and carbon dioxide. Good for making explosions and fires. Not good for the nose to smell, for the atmosphere, for human health. Very good if it is used to make electricity. For example, a generator plant at Wellington's Southern Landfill has an internal combustion engine designed to run on landfill gas. It produces around 8 GWh of electricity per year.

Case Study Waste 2 Gold

In 2010 the New Zealand government announced that SCION, a Crown Research Institute, was to get a $1 million grant. SCION works to get sustainable design which results in the lowest possible environmental footprint. It also works to get bioproduct development and new bioprocesses from renewable resources. The grant was to transform sludge from Rotorua's sewers into useful products. It involved SCION and Rotorua District Council joining forces to develop a new approach to the management of waste.

The project was called 'Waste 2 Gold'. Rotorua has about 20 tonnes of sludge going to a landfill each day. Waste 2 Gold uses new technology developed by SCION to turn this biosolid waste (sewage sludge) from Rotorua's wastewater treatment plant into products such as energy and fertilisers. It 'cooks' the biosolids and breaks them into reusables. For example, methane can be used to make electricity. This is a sustainable solution to the disposal of solid organic waste and a good way to reduce greenhouse gas emissions.

The global market for bioplastics is expected to take off. Bioplastics are a form of plastic made from renewable sources, such as vegetable oil and cornstarch, instead of the old plastics made from non-renewable petroleum. SCION is turning kiwifruit waste into compostable bioplastics. It makes combined spoon and knife utensils known as 'spifes' to be included in kiwifruit packages sold overseas by Zespri. Another innovation is the biopeg. This is a tent peg that can be left in the ground to decompose harmlessly.

ISBN: 978-0170210348

Case Study Space Junk

The problem

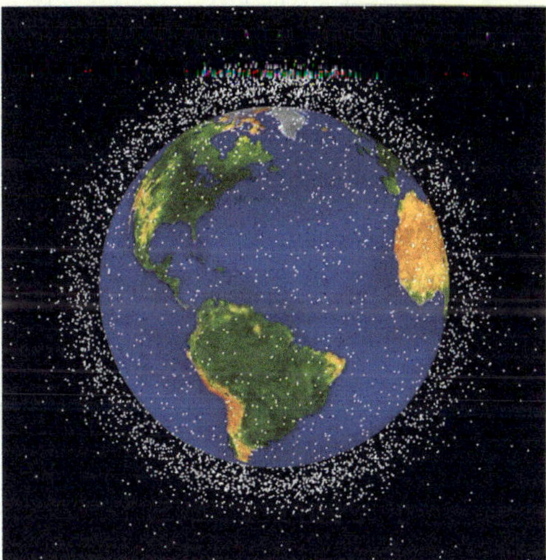

This is a computer-generated image of objects in Earth's orbit that are currently being tracked. Approximately 95 percent of the objects in this illustration are orbital debris and not functioning satellites. The dots represent the present location of each object.

- Up in space more than a thousand orbiting satellites, belonging to more than 40 countries, power a big number of services for Earth's people, ranging from cellphones to banking, weather reports to navigation, mapping to communication. Humans use space. Humans create waste. Therefore humans have created waste in space.
- This waste in space is called space junk. There are dead satellites, dust from solid rocket motors, paint flakes, garbage bags, things space walkers have lost such as bolts and cameras.
- Hundreds of thousands of bits of space junk zoom round at speeds that can reach many thousands of kilometres per hour.
- Crashes with space junk threaten astronauts, spacecraft, operating satellites and the services they give to Earth. Small bits can puncture solar panels on spacecraft. Larger bits can destroy spacecraft. A piece of metal space junk the size of a tennis ball is as dangerous as 25 sticks of dynamite. Even a tiny flake of paint from a satellite can make a dent in a space shuttle window. Larger junk bits could reach ground. They might hit something or someone. They might have dangerous chemicals in them. Wreckage from a Russian spy satellite once passed close to a plane travelling between Auckland and Chile with 270 passengers.

The solution?

- Countries stop treating space as a place where there are no or few rules.
- Get scientists and governments to agree that the use of space should not threaten the ability of future generations to use it as well.
- The United Nations made guidelines in 2008 to get countries to use space in a sustainable way. It should make countries stick to these guidelines.
- Build better controlled self-destruction into satellites.
- Use new technology such as powerful lasers to clean up junk already there.
- Get robot missions to dock with dead satellites and fire rockets either to boost them into graveyard orbits or deorbit them so they crash into the sea.

Can you see a possible problem with inventing ways to kill space junk? Satellites are important in modern warfare. They spy on battlefields and enemies, even allies. They provide communication links. Having the technology to knock them out would be useful.

ISBN: 978-0170210348

1 Describe what is meant by the saying 'Somebody's rubbish is another person's treasure.' Then outline the advantages and disadvantages of second-hand shops.

2 Disposable nappies often feature in discussions about landfills. A baby wearing them puts two tonnes of waste into a landfill and that may take up to 500 years to break down. Make a poster suitable for a booklet for new mothers about the advantages and disadvantages of disposable nappies.

3 Which group can come up with the longest list for reusing children's drawings, plastic bags, broken crockery, egg cartons, magazines.

4 Read the following and answer the questions.

A recent issue of the fashion magazine Vogue featured Prince Charles doing his bit for fashion, and the environment. Wear wool rather than man-made fabrics, he said. Prince Charles has a passion for reusing things and repairing them. He has had leather from an 18th century wreck made into a pair of shoes for himself. He says vintage clothes are good for the fashion industry and consumers because they avoid waste and save resources.

- **a** Describe the picture you would ask the artist to draw to go with the article in Vogue if you had been the editor.
- **b** What are vintage clothes?
- **c** Give proof that shows Prince Charles is environmentally-friendly.

5 Describe what space junk is, why it is a problem and what methods you would use to deal with it if you were the official in charge.

6 Find out what happens to the waste collected in your area.

ISBN: 978-0170210348

Sustainable Species

species = type or kind of plant or animal

biodiversity = the variety of plants and animals

ecosystem = an environmental area, such as a lake, and all the plants and animals in it

habitat = place where a plant or animal lives

native = naturally from a place, opposite of introduced which means brought in from another country

indigenous = native

endemic = native

extinct = the last member of a species dies

alien = non-native

New Zealand is one of the last places on Earth that humans settled. This means people have had less time to have an impact on the biodiversity here. Yet New Zealanders have one of the worst records of killing off native species, especially bird species. People didn't realise. They wrecked habitats by lighting fires, chopping down trees and draining wetlands. They over-fished, over-hunted and over-gathered. They introduced new animals and plants that went to war with the native ones. The extinct species would have been interesting to know. The Haast's eagle, for example, was the biggest eagle in the world. It was so big it probably ate moa. Human bones have also been found in its nest sites.

New Zealand is not the only country to have had extinctions. The extinct dodo, for example, was a goose-like bird on the island of Mauritius off Africa. It was clumsy and couldn't fly. Its name came from a Portuguese word meaning a simpleton. Today there is a saying that you can be as dead or as stupid as a dodo.

ISBN: 978-0170210348

Scientists say there were five great mass extinctions in the past. For example, Tyrannosaurus Rex was a victim of the fifth mass extinction. Now we are in the sixth extinction. This extinction is different from all the others. That's because humans are causing it. It began thousands of years ago. Scientists say we know less about Earth than we do about Mars. There are vast numbers of life forms still undiscovered. Yet Earth is losing an average of one species every 20 minutes. Thousands of species become extinct before scientists have a chance to describe them. That's not sustainability in action.

Nowadays, government in New Zealand works to keep biodiversity healthy. It gives money to landowners and community groups for projects aimed at looking after biodiversity. It has a Department of Conservation and NIWA (the National Institute for Water and Atmospheric Research). There are many other organisations such as the Kiwi Recovery Programme and the Project Crimson Trust for pohutukawa.

International groups also look after biodiversity. For example, marine biologists are worried that wild seahorses may face extinction. Seahorses are threatened because they are traded on the black market, made into Asian medicine, killed by pollution, and are losing their coastal habitat. Biologists have started to breed them in aquariums. This is time-consuming because the creatures need three feedings a day.

Scientists say biological invasions are a big threat to the sustainability of species. On the list of the world's worst 100 alien species are many that you will recognise because they have been introduced to New Zealand - myna bird, Asian tiger mosquito, Asian gypsy moth, Sea Star, Dutch Elm disease, goat, red deer, carp, house mouse, stoat, pig, gorse, brush tail possum. Governments set up groups to stop foreign plants and animals coming into their countries.

ISBN: 978-0170210348

Case Study The Kiwi

You wouldn't want to be the person who let a dog run wild in a forest and saw it kill the last kiwi. The kiwi has great value. It's a sign of identity for New Zealanders who are even called Kiwis. It's linked to the clean and green image that supports exports and tourism. If your children are ever to see a kiwi, the species has to be saved from extinction now. The Department of Conservation, the Royal Forest and Bird Society, and the Bank of New Zealand launched the Kiwi Recovery Programme in 1991. It works to protect kiwi chicks and eggs from dogs, cats, ferrets, stoats and weasels. It raises kiwi chicks in captivity and releases them into the wild when they can defend themselves against enemies. It keeps up its research into kiwis, and gets help from the community especially in places where kiwis are on private land. It helps set up safe places for kiwis to live and breed.

Snoopy was a Brown kiwi. He had lost a leg in a possum trap and lived with a family from the time he was three months old. He helped educate people about kiwis. He was taken to schools and never minded being in a noisy classroom. He died at age 15. Snoopy's educating role has been taken over by Sparky. His leg was so badly mangled in a trap that it had to be amputated. Sparky's minders say he hops about like a kid on a pogo stick. If you were holding Sparky and you dropped him, he'd die because kiwis have no breastbone.

1. Learn the spellings of the terms in the box on page 53.
2. Give three words that could describe the relationship of the kiwi to New Zealand.
3. Design a home for Sparky.
4. Make a cartoon strip about a native plant or animal teaching another one how to avoid extinction.
5. Find out what the World Wildlife Fund is and what it does.
6. Find out about some extinct or near-extinct species from around the world.

ISBN: 978-0170210348

The Water Footprint

A water footprint shows the total volume of water used to produce goods (things you can see such as food and clothes) and services (things you can't see such as actions and work) consumed by a country, community, business or person. It includes:

- blue water – fresh water in lakes, rivers, aquifers
- green water – water from rainwater stored in the soil
- grey water – polluted water.

Being sustainable means using blue water wisely and not making grey water. Humans have polluted much water. Some rivers have so much rubbish in places that the only way you can tell they are rivers is seeing a boat pushing its way through the rubbish. If you were in that boat you wouldn't want to fall out. The water under the rubbish has chemicals from factories and toilets. It looks like thick oil. Lake Karachay in Russia usually wins the title for most polluted water on the planet. It was the dumping site for radioactive waste. Would you want to picnic on its banks?

New Zealand is lucky when it comes to water. It has a lot of blue water. So are Canada, Scandinavia, Brazil, Britain and parts of Russia. Many other places such as Australia face water shortages. There are tensions between countries over water. Would Israel and Jordan go to war over it? Would India and Pakistan? Would Singapore and Malaysia?

ISBN: 978-0170210348

The Coca-Cola Company operates over a thousand manufacturing plants in about 200 countries. Making its drink uses a lot of water. Critics say its water footprint has been large. Coca-Cola has started to look at its water sustainability. It has now set out goals to reduce its water footprint such as treating the water it uses so it goes back into the environment in a clean state. Another goal is to find sustainable sources for the raw materials it uses in its drinks, such as sugarcane, oranges and corn. By making its water footprint better, the company can reduce costs, improve the environment, and benefit the communities in which it operates.

The water footprint of a country is related to what its people eat. Eating a lot of meat means a large water footprint. The more food comes from irrigated land, the larger is the water footprint.

The Water Footprint Network is an international group. It includes governments and organisations. It calculates water footprints for countries. For example, the water footprint of New Zealand shows the water used to produce the goods and services consumed by New Zealanders. This is not just what water is used in New Zealand. It includes any water used in other countries. The US has the largest water footprint.

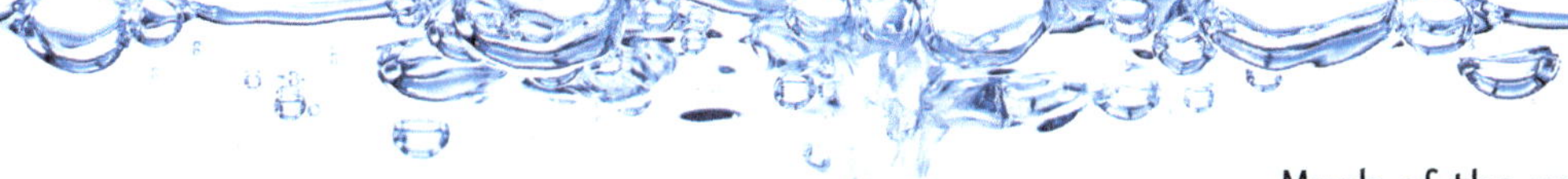

Examples of global averages

Product	Water needed to make it (litres)
1 kg wheat	1350
1 kg maize	900
1 kg rice	3000
1 kg beef	16000
1 glass of milk	200
1 cup tea	35
1 slice bread	40
1 pair leather shoes	8000
1 microchip	32
1 cotton t-shirt	4000

Much of the water use is hidden. For example, you'd think the water involved in a cup of coffee is just the water in the cup. But actually 140 litres of water is involved. That's how much water was used to grow, produce, package and ship the coffee beans. A hamburger needs an estimated 2,400 litres of water. The technical term for this hidden water is virtual water. Each American consumes around 6,800 litres of virtual water every day. That's more than three times what a Chinese person consumes.

An Australian water conservation specialist did a study on a jar of pasta sauce and a bag of peanut M&M's. The water needed to grow tomatoes, sugar, garlic and onions for the sauce was 197 litres. For the M&M's, cocoa and peanuts it was 1,135 litres. End of story? No. Tomato plants are often grown in hot, dry climates. They get irrigation water from the same locations as human drinking water. Cocoa and peanuts in M&M's are grown where crops get rainwater from the ground. The specialist calculated that the pasta sauce was about 10 times more likely than the M&M's to add to water scarcity. This was because the blue water used for tomatoes was much higher stress-weighted than the green water for the cocoa and peanuts. Blue water is stress-weighted because it is consumed at rates faster than its short term replacement.

ISBN: 978-0170210348

Case Study Little and Big Problems to Solve

In a water-rich country like New Zealand everyone can save water. Turn the tap off when you brush your teeth for a minute and you save more than a bucketful of water. Put a gizmo in your toilet and you save thousands of litres a year. The gizmo is a simple lead weight; it takes less than a minute to hang it inside the cistern tube. As soon as you take your finger off the flushing button the gizmo stops the flushing. So you choose how much water you need to use.

In a country with a big population, problems to do with water are taking a while to solve. In China, dams and irrigation have turned some rivers into trickles. Factories pour untreated waste and chemicals into rivers. 90 percent of groundwater for cities and 75 percent of rivers and lakes are polluted. 700 million people drink polluted water every day. North China now relies on underground water formed 10 thousand years ago. South China has more water but much is polluted. The World Bank warns that if this water situation keeps going, by 2020 there will be 30 million environmental refugees. Government plans to reduce water pollution by using wastewater treatment. It is also putting money into restoring wetlands. China, like some other countries such as Malta, Israel and Singapore, has some desalination plants. These remove salt and other minerals from ocean water to make it safe to drink. They are expensive to build, use a lot of energy, produce greenhouse gases, can kill fish when the salt water is taken in, and produce a concentrated salt stream that can pollute when dumped back into oceans.

1. Is water, after air, the most important thing on the planet? Give reasons for your answer.
2. Is it possible that scientists will invent something to replace water? Give reasons for your answer.
3. Have a class competition among groups for the longest list of ways to save water.
4. Discuss the idea that food products should have a label to show how much water was used to produce them.
5. Explain how finding a water footprint is a complicated process.
6. Find out if New Zealand and Australia are generally regarded as water-rich or water-poor countries.

ISBN: 978-0170210348

Wetlands - Sustainability's Friends

Wetlands are where water and land meet, making 'wet land'. These shallow ponds and marshy areas covered with vegetation such as reeds are excellent resources to have. Yet people haven't always been kind to wetlands. For example, New Zealanders have drained or filled in over 90 percent of their wetlands.

Today, wetlands are making a comeback. The United Nation's International Year of Biodiversity in 2010 said wetlands were brilliant friends to biodiversity and enemies to climate change. The World Wetlands Day is held on February 2 to celebrate wetlands.

They make habitats for hundreds of species of birds, fish, insects, mammals.

They act as giant kidneys by filtering out pollutants before they let water leave. Plant roots and stems trap and gather the pollutants.

They are the most biologically diverse of all ecosystems. That means they have more plants and animals than any other type of natural environment.

They catch silt (mud or clay in water).

They are good for tourism and recreation such as bird watching, walking, photographing, whitebaiting.

They can remove and store in their soil, greenhouse gases from Earth's atmosphere.

They act like giant sponges to:

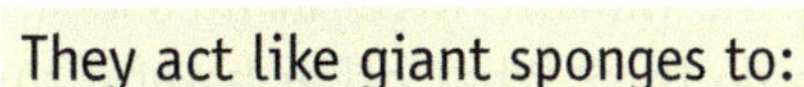

- absorb huge amounts of water in a short time
- slow the flow of surface water and reduce the impact of floods
- store water
- prevent soil erosion.

ISBN: 978-0170210348

Case Study Whangamarino Wetland

Whangamarino wetland is recognised as being internationally important. It's a lot smaller than it was before Europeans arrived but DoC (Department of Conservation) has worked to control introduced willows and gorse, bring back native plants, and get rid of pest fish such as catfish, pest plants such as alligator weed and pest animals such as possums and mustelids (ferrets, stoats, weasels).

1 Why would visitors to a wetland be asked to make sure their footwear and gear is clean?

2 If Whangamarino wetland did not exist, Council would have to build stopbanks along the Waikato River at a cost of many millions of dollars. Why would it have to do this? What other helpful things would the wetland be doing for the area?

3 From which lake and in which direction does the Whangamarino River flow?

4 What are the main types of vegetation in the Whangamarino Wetland?

5 What methods of travel through the wetland could you use?

6 Find out what Ramsar Status is and which wetlands in New Zealand have it.

ISBN: 978-0170210348

Chemical Crunch-Time

THE GOOD
EXAMPLE: The chemical industry has given humans more control over their own bodies and over their environment.

THE BAD
EXAMPLE: The Fruitgrowers Chemical Company built a pesticide plant at Mapua, Nelson in 1932, and later began producing organochlorine pesticides. Chemicals got into the soil, groundwater and nearby estuary (where ocean and river meet). The plant closed in 1988. In 1999 the government decided to help Tasman District Council with its cleanup. In 2010 the government was still reporting on its investigations into the site.

THE UGLY
Recently reported in the US: Researchers in New Zealand found that female snails exposed to the chemical TBT were growing penises from their heads.

pesticide = chemicals used to kill pests such as rodents and insects

insecticide = pesticide used to kill insects

herbicide = chemicals used to kill weeds

The chemical barrage has been hurled at the fabric of life.

Rachel Carson was an American biologist and ecologist. She worried about the use of synthetic chemical pesticides after World War Two. In 1962 her book *Silent Spring* warned the public about misusing pesticides. The chemical industry attacked her but she insisted on the need to protect human health and the environment. Many people say she did more than any other for the environmental movement around the world. She could have been called Mrs Sustainability.

Each chemical preparation has the possibility to be good, bad, or ugly. Lindane, for example, is an organochlorine pesticide used to treat head lice in people. Now 50 countries have banned it because it can attack the nervous system. For every amount of Lindane produced, six to 10 amounts of waste are produced.

ISBN: 978-0170210348

What environmentalists say	What the chemical industry says
The chemical industry needs to: • reduce its use of nonrenewable energy and resources • stop using so much water • reduce its greenhouse gas emissions • stop making toxic products and waste • make products that are recyclable and biodegradable • stop being so secretive.	The global chemical industry has set up Responsible Care. Most of the world's big chemical manufacturers have signed up to it. It's the chemical industry's global organisation that aims to get sustainable development. It's about helping the industry to operate safely, and with care for future generations. It is trying to get companies to be open and honest about their actions.

Read the following and answer the questions that follow.

DDT was the first of the chlorinated organic insecticides. It does not occur naturally. The man who discovered its effectiveness won the Nobel Prize. After World War Two its use increased because it worked against the mosquito that spreads malaria and the lice that carries typhus. It saved millions of lives. It was cheap to make. It was also used to control grass grub and porina moth.

However, problems related to its use began to show. Many species of insects developed resistance to it. It was toxic for fish, eels and birds. It thinned bird eggshells. Fish ate poisoned worms; birds, sea lions and people ate the fish. There were suspected links to diabetes and cancer in humans. It can sit in soil for years. Scientists find DDT in soil, water, and even in ice from the Arctic to Antarctica. The New Zealand government has signed and ratified the Stockholm Convention which controls the making and use of persistent organic pesticides, which includes DDT. Many countries, including New Zealand, have banned DDT. Some people say it should never have been banned anywhere.

1. What is DDT and why was it used?
2. Describe what Rachel Carson might have said about the banning of DDT.
3. What evidence could you use to back up this statement: A blanket of DDT covers the earth.
4. Make a labelled drawing to show how DDT can move into the food chain.
5. Explain why banning DDT is an issue.
6. The world's worst industrial accident was at a pesticide plant in Bhopal in India. Find out what happened, when and the results.

ISBN: 978-0170210348

The Ecological Footprint

It hasn't been around all that long. But now scientists, governments, groups and individuals use it to measure sustainable development.

Every person on the planet has one. It's all about your overall impact. You can go online and answer quizzes to get an estimation of your ecological footprint.

You need a certain amount of land and water to provide the resources you consume and to deal with the waste you make. The total amount is your ecological footprint. It's like all your footprints, such as carbon footprint and water footprint, rolled up into one giant footprint.

It now takes the Earth one year and six months to regenerate (renew) what humans use in a year. That's like saying we need another Earth, or at least half of it.

Think of all the calculations needed to measure ecological footprints. It takes several years to get all the data together. Calculations use a global hectare as a standardised hectare of land able to produce resources and absorb wastes at world average levels. Say the world has an average 2.1 ha available per person. If the average ecological footprint is higher than 2.1 ha, Earth and humans are in trouble. Has it shot over 2.1 ha? Yes. This is called ecological overshoot.

The United Arab Emirates, the US, Kuwait, Denmark and Australia have the biggest ecological footprints. New Zealand isn't far behind. In the recent World Wildlife Fund's report New Zealand had moved from needing 5.9 global hectares per person to an average of 7.7 global hectares. It showed New Zealanders used too much of natural resources. If demand kept growing at the same rate, two planets would be needed in the mid-2030s - or 3.5 planets if everyone on Earth used resources at the same pace as New Zealanders.

ISBN: 978-0170210348

The Bad News

The World Wildlife Fund's report said the biggest human stress on the planet was from using fossil fuel. In New Zealand the main growth in carbon emissions since the ecological footprint came on the scene is energy. New Zealand is now second only to the United States both in the number of cars owned per person and in the number of kilometres travelled in those cars.

The Good News

Ecological footprints help Governments know what laws to pass. They can lessen the impact of human actions on the environment. New Zealand has a Resource Management Act. It works to get sustainable management of resources. Councils have to prepare plans to show how they will manage their environment. They have to think about what impact a proposed action could have on the environment.

Case Study New Zealand Icebreaker

Icebreaker is a Kiwi clothing firm. It's part of the growing movement in the fashion industry to reduce its ecological footprint. Making clothes out of materials that can be traced back to where they came from means you can see how sustainably produced they were.

Icebreaker makes clothes from merino wool that it says is 'the world's most exclusive natural fibre, grown in the pristine Alps of New Zealand.' Some garments are blended with possum fur. Sustainability is Icebreaker's big thing. Each Icebreaker item comes with a code. It's called the Baa-code. You, as the consumer, can punch the code into a computer to trace the garment through every step of the production process. This is a first for the clothing industry. It begins with the sheep station or stations in New Zealand from which your wool came. You can find out about each - how many sheep it has, how many hectares it has, how long the farmers have been there, how much wool they produce.

ISBN: 978-0170210348

Ecological overshoot impacts on future generations.

1 What is ecological overshoot and how are humans dealing with it?

2 Find evidence for this statement: More than three-quarters of the world's population lives in countries whose consumption outstrips environmental renewal.

3 Name 20 actions you could take to reduce your ecological footprint.

4 Explain why figures for the ecological footprints of countries are not up-to-date.

5 Experts say New Zealand must reduce its greenhouse gas emissions. What evidence is there to back up the experts' opinion?

6 What other measures can the fashion industry take to make it more sustainable?

ISBN: 978-0170210348

Fishy Issues

Issue 1: Overfishing
Overfishing is unsustainable fishing. People catch fish at a faster rate than the fish can reproduce. Result? Fewer fish. Two men in a rowing boat 100 years ago could catch more fish than some high-tech fishing trawlers do today.

Issue 2: Illegal, unreported, and unregulated (IUU) fishing
Governments and organisations make laws about fishing to make sure fishing is sustainable. Unsustainable fishing is when people break the laws, make false reports about their fishing catches and use pirate boats to escape being caught.

Issue 3: Bottom trawling
The trawler drags a fishing net, called a trawl, along the ocean floor. It grabs at everything in its path. Not just fish but things like reefs and corals where fish live. It stirs up the bottom where some toxins such as DDT have settled. This brings the toxins back into the food chain. Critics say it damages seamounts where some fish gather to feed or mate.

Issue 4: Bycatch
This is what fishing gear catches by accident. It includes other fish types, and animals such as birds, sea lions, fur seals, dolphins, turtles.

Issue 5: Pests
These are plants and animals that hitch rides on vessels and arrive in a new place. There they terrorise the locals by attacking them or taking over their homes. For example, the Northern Pacific Sea Star eats mussels. The Chinese Mitten Crab steals food and shelter from local crabs. It likes burrowing so it damages banks and clogs drainage systems. The Sea Squirt, which does squirt seawater, steals space and food from mussels and oysters.

ISBN: 978-0170210348

Issue 6: Ghost fishing
When fishing people lose or throw away into the ocean gear such as pots and nets, that gear can keep catching and killing fish.

Issue 7: Dynamite and poisons
In some places people dive into the sea and spray cyanide into coral reefs. That poisons the coral as well as the fish. Then they collect the stunned fish and put them in fresh water to rinse. This kills off so many fish that more fish have to be caught. Dynamite fishing (blast fishing) stuns or kills schools of fish and destroys their habitat such as coral reefs.

Issue 8: Pollution
Humans continue to use the ocean as a dumping ground. Industrial waste, sewage, rubbish from ships. Chemicals that cause eutrophication - increased nutrients which causes too much plant growth and decay. Toxins that don't break down, such as radioactive waste and mercury. Plastic, especially plastic bags. Oil spills, ballast water held in tanks to make vessels more stable. Even noise – from ships, sonar, oil exploration, deep sea mining. Out of sight, out of mind, people used to think. Except now they can see it. In 2010 a massive oil spill in the Gulf of Mexico had scientists talking about an underwater blizzard of gooey stuff often called sea snot.

Issue 9: Greenhouse gases
Up into the atmosphere goes pollution such as exhaust fumes and refrigerant gases from fishing vessels and processing plants.

Issue 10: Waste fish
How would you like to be a shark hunted to make shark fin soup? You get hooked and reeled in. Someone cuts your fins off. You're still alive. That someone chucks you back into the ocean. Oops, no fins mean no swim. You flounder about for hours before you suffocate and die. If you're lucky, other sharks will come and eat you first.

ISBN: 978-0170210348

Scientists and environmentalists say you can help by supporting these solutions.

Make rules to stop overfishing.

Change the design of longlines so they don't catch birds.

Stop councils and governments letting untreated waste go into oceans.

Make rules to stop the dumping of gear at sea and fine anyone caught doing this.

Stop buying fish if it comes from unsustainable fishing practices.

Find a better fishing method than trawling.

Pick up gear lost at sea such as gillnets lying on the bottom.

Set up zones to keep trawlers away from habitats they could damage.

Stop waste such as shark bodies.

Stop dynamite and cyanide fishing.

Make all fisheries have an environmental impact assessment.

Use patrol boats.

Invent technologies to improve fishing such as traps that collapse and biodegradable fishing gear.

Improve net holes to stop capturing baby fish who never have a chance to reproduce.

Don't buy souvenirs such as polished shells, jewellery made from tortoiseshell, pictures with dried seahorses in them, ashtrays made from clams.

Put a fisheries technician on every boat to collect information and send it digitally to a database so fishing industries can manage their fisheries in real time.

Make every fishing country have a quota system. For example, the government and the fishing industry in New Zealand spent years planning a Quota Management system. Fishers are granted a right to take a certain amount of a certain species of fish. This right is called a quota. Each year scientists and the fishing industry work out how much quotas, in kilograms, should be. This means they have to first estimate how many fish exist and how fast they reproduce. Then they have to work out what is a sustainable amount that can be taken. Fishers can buy, sell, trade or lease quotas.

Case Study Hoki

- Has other names such as blue grenadiers, blue hake, whiptails, whiptail hake, New Zealand whiting.
- Grows very fast.
- Found off the coasts of New Zealand.
- Flesh is white and rich in Omega 3.
- Reproduces in large numbers.
- Big eyes make it look frightened.
- Certified as sustainable.
- Has never been overfished.
- Government monitors stocks to keep it sustainable.
- Exports of it earn a lot of money for New Zealand.

ISBN: 978-0170210348

However...
Recently Waitrose, a big British supermarket chain, refused to stock New Zealand-caught hoki. Its reason was that the industry used bottom trawling. (Other buyers also have concerns about the bycatch of New Zealand fur seals, albatrosses, petrels, and basking sharks.) Soon after, Trader Joes, a US chain with hundreds of supermarkets, stopped selling New Zealand's orange roughy based on customer feedback and in support of work to get sustainable seafood. Canada's largest retail chain, Loblaw, also stopped stocking New Zealand's orange roughy for the same reason.

1 Make some labelled sketches to show what the following mean: overfishing, pirate boat, trawl, seamount, Sea Squirt, bycatch, ghost fishing, blast fishing, sea snot, noise pollution, refrigerant gases, Waitrose, petrel, gillnet, quota.

2 You are the Minister in charge of New Zealand Fisheries. You are to deliver a talk to an international conference on sustainable fishing. Make a list of topics you would like your staff to research to help you prepare your talk.

3 Use hoki and orange roughy as examples to explain why sustainability can mean more than one thing.

4 Vox-pop stands for vox-populi which is Latin for voice of the people. It is the opinions of people on an issue. Sometimes photos of people appear beside their opinions. Make a vox-pop on the issue of bottom trawling. Try for at least five opinions.

5 In your group, decide how you would go about setting a quota for a fish of your choice.

6 Choose one fish caught in large numbers in New Zealand. Prepare a profile of it and give it a rating out of 10 for sustainability.

ISBN: 978-0170210348

Trees - Sustainability's Main Mates

Wood

Wood is New Zealand's most sustainable raw material. Look how sustainable it is in comparison to some other building materials.

During production (one tonne)		
Building material (1 tonne)	Carbon dioxide absorbed from the atmosphere	Carbon dioxide emitted into the atmosphere
concrete	none	159 tonnes
steel	none	1.24 tonnes
aluminium	none	9.3 tonnes
wood	1.7 tonnes	none

As a tree gets bigger, the amount of carbon it has also grows. This is because wood is almost 50 percent carbon by weight. For example, if you had a piece of furniture handed down the generations from 1500, it would still hold the carbon fixed in it from that time.

cutting trees down (felling trees) = deforestation

planting trees = afforestation

The cycle

Somebody plants a tree or a seedling appears.

The tree grows, absorbs carbon dioxide and stores it as wood.

The tree is turned into timber for building. Its stored carbon dioxide stays in it.

The wood rots or gets burned as fuel. The carbon dioxide is released.

ISBN: 978-0170210348

It makes sense to:

- have sustainable forestry. That's one that is managed so that as trees are felled they are replaced with seedlings that grow into trees. That way people can plant, grow, cut and replant trees forever, always being friendly to the environment and fighting climate change.
- control climate change because it is not always kind to forests. Australia, for example, loses forests to drought and fires.

Economic development and deforestation are closely linked

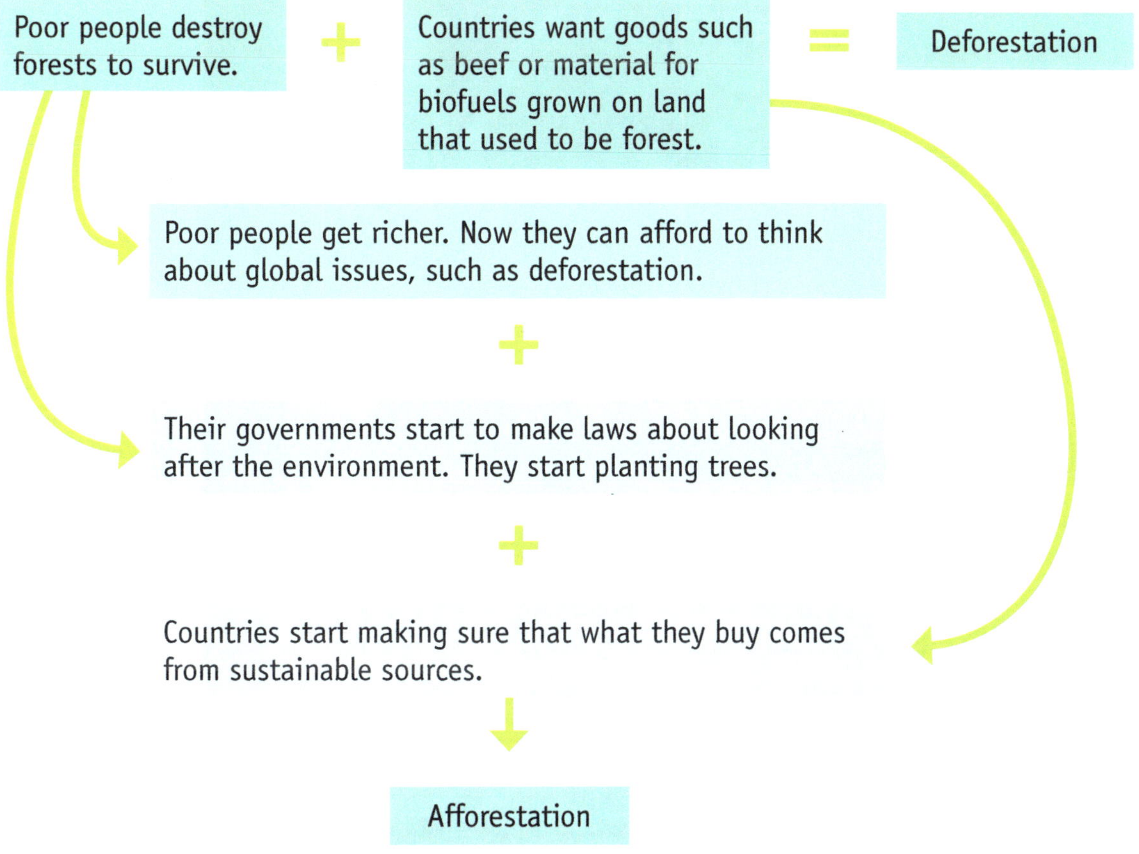

For example, the United Nations records show that over the recent 10 years, nearly eight million hectares of forest a year were allowed to re-grow or were planted again.

On the global scene, average wood demand increases by three tonnes every second.

= Great news for New Zealand's economy. New Zealand is one of the few countries in the world that can increase its wood production AND keep its forestry sustainable.

ISBN: 978-0170210348

UN-REDD

PROGRAMME

Act now!

Time is running out. It is now clear that in order to constrain the impacts of climate change within limits that society will reasonably be able to tolerate, the global average temperatures must be stabilized within two degrees Celsius. Each month, over one million hectares of tropical forest are destroyed, releasing hundreds of millions tons of carbon dioxide into the atmosphere. We need to act now to reduce deforestation and the degradation of forest land to cut emissions and avoid the dangerous consequences of climate change.

The UN-REDD Programme is a partnership between FAO, UNDP and UNEP. It supports developing country efforts to reduce emissions from deforestation and forest degradation - commonly known as "REDD".

The main international effort to speed up the change from deforestation to afforestation is REDD (Reducing Emissions from Deforestation and Forest Degradation). It pays people in poorer countries to leave trees standing. This depends on countries with important forests such as tropical rainforest having a decent government, so it does not steal the money and use it for something else.

Early settlers chopped down much of New Zealand's kauri forest for wood. That's illegal now as kauri forestry is not sustainable. Kauri trees take a very long time to grow. Plantation forestry is sustainable. Its trees are grown especially to produce timber. They are usually ready for harvest after 25 to 30 years. The New Zealand forest industry isn't perfect. Critics point out it clear-cuts (cuts down a whole area in comparison to selective logging which leaves some trees standing), and it uses herbicides and insecticides.

ISBN: 978-0170210348

Case Study Pinus Radiata

Radiata pine (Pinus radiata), is known as Monterey pine in its place of origin, California. It makes up nearly 90 percent of New Zealand's sustainable plantation forest. The pine grows about seven times faster here than in the US and 20 times faster than in Canada. New Zealand, along with Chile and Australia, are the top growers of this species worldwide. It grows well throughout New Zealand on a variety of soil types, including coastal sands, heavy clays, gravels and volcanic ash deposits. It is the main reason New Zealand's industrial wood supply has changed almost completely from natural forests to managed forest plantations. This sustainable forestry is set to double by 2025. Kaingaroa Forest in New Zealand is the biggest plantation forest in the southern hemisphere.

1 Which is the only building material that absorbs carbon dioxide from the atmosphere when it is produced rather than emitting more?

2 Find proof for this statement: Planting more forests means creating a bigger carbon sink.

3 Explain the difference between trees in a plantation forest and coal.

4 Use the following facts about New Zealand forestry to help you write a paragraph about the importance of it to the economy.

Forests = 8.2m hectares, 30%, of land. 6.4m hectares = indigenous, 1.8m = plantation forests. Pine = 89% of total plantation, 7% of land use. Douglas-fir = 6% of total population.
New Zealand = 0.05% of world's forest resource, supplies nearly 9% of Asia-Pacific forest products trade. In 30 years, 1ha of NZ pine can produce as much quality wood as 10ha of tropical forest in Southeast Asia, or 40ha of Amazon forest.

5 Why might a sustainable forest have trees of all ages and maybe even different species?

6 Find out how photosynthesis works and draw a diagram to show the process.

ISBN: 978-0170210348

Amazon Rainforest - From Tree-Choppers to Tree-Huggers

For years you've probably heard about the Amazon rainforest and how it's been disappearing at a most unsustainable rate. Why do we need to be worried about that?

Reason 1: It's the world's greatest remaining natural resource. It's 54 percent of the total rainforests left on Earth. It covers parts of nine countries. It's bigger than most countries.

Reason 2: It's the most biodiverse natural thing on the planet. A single rainforest reserve in Peru is home to more species of birds than are found in the whole US. There are more fish species in the Amazon than are found in the whole Atlantic Ocean.

Reason 3: Only a tiny fraction of its plant and animals species have been seen by humans and even fewer recorded by scientists.

Reason 4: It's called the lungs of the planet because it continuously recycles carbon dioxide into oxygen. It produces probably more than 20 percent of Earth's oxygen. It sucks in so much carbon dioxide that it's the biggest factor fighting global warming. If the Amazon went up in smoke, it would pour out more than a decade's worth of fossil-fuel emissions. Earth's atmosphere would change if the forest was destroyed; it could reduce rainfall across the Americas.

The good news for sustainability is there is hope for the Amazon rainforest. Official figures in 2010 showed deforestation had declined greatly compared to 2009. Satellite data to spot the felling of trees has allowed people to better track and stop illegal loggers, who can no longer hide under cloud cover. Many tree-choppers have become tree-huggers.

ISBN: 978-

As usual, the economy (money) is involved. It causes the problem of deforestation but it can also fix it.

People cut down the forest to make a living eg. make a cattle ranch which makes the land worth $60 per acre, or sell the timber which makes the land worth $400 per acre.

Show people they can make a living from the forest without chopping it down - sustainably harvest medicinal plants, fruits, nuts, oils, and other resources like rubber, chocolate, and chicle (used to make chewing gums) which makes the land worth $2,400 per acre.

For example, Indians of Pukanu village made a deal with British seller Body Shop to export Brazil nut oil. People gather nuts in the forest. They bring them back to a factory with traditional mud walls and straw roof. The nuts dry in the sun and then people remove the oil.

Case Study STARO

Save the Amazon Rainforest Organisation (STARO) was set up to protect the Amazon Rainforest. It helps the people living there to find ways of earning money without cutting down trees. That is sustainable use. An example is Buy the Bees. This project buys native stingless bees for local people to make honey for sale. The stingless bees depend on the rainforest for survival. Unlike honey bees, they can't change to survive among crops. If the rainforest goes, so will the bees. Therefore stingless bees can teach forest people about the importance of their forest. Making honey is sustainable and cheap.

Another STARO project is 'Alley Cropping'. Rainforest soil where the trees have been chopped down loses its goodness. Alley cropping plants Inga trees in rows that form alleys. The Inga is a native that can survive in the soil. Its thick leaves fall on the soil after pruning and make a mulch in the alleys. Crops are grown in the mulch which feeds the crops. This cycle can be repeated over and over again. People can grow their basic crops and also cash crops. Before this, the people would have to keep cutting down trees to make new plots because the soil lost its goodness. Often their plots would be a long way from their homes. With alley cropping, the same plot is used over and over. It can be close to home which means the people can guard it.

ISBN: 978-0170210348

1 Would you be interested in visiting the Amazon rainforest? Give reasons for your answer.

2 1,500 km^2 of rainforest were ripped down between August 2009 and May 2010, compared with 3,000 km^2 in the same period the year before. Give as many reasons as you can why that is good news for the world.

3 James Cameron is the famous movie director of *Titanic* and *Avatar*. He and his wife live on a solar-powered ranch. This is useful publicity for solar power. If you were a celebrity who wanted to help save rainforest, what sort of things could you do?

4 Make a diagram to show how alley cropping works. Include the benefits of it on your diagram.

5 Explain why sustainable resources - not the trees - are the true wealth of the rainforest.

6 Find out how and why the term tree-hugger is changing its image.

ISBN: 978-0170210348

Food Miles

When countries want to buy products from New Zealand, Kiwis have to put those products on to planes or ships. They can't do what many other countries can – load a truck and drive over a border. 'Ooh,' say some people. 'Kiwis are racking up too many food miles.' These people won't touch food that is not local. They call themselves locavores.

Locavores talk a lot about food miles. That's the distance food travels between where it's made and where it's eaten. Because its inventor was from the United Kingdom (UK), which measures in miles and not kilometres, the term is food *miles*.

How ideas change

Old Idea Food that travels distances to consumers uses more energy, emits more greenhouse gases, and damages the environment more than food produced locally.
New Idea Nonsense, say many experts. Food miles are useless for measuring sustainability. They don't tell you anything except how far the food has travelled.

These are four examples used by experts to show how measuring food miles is not always helpful.

Example 1

A recent study compared some New Zealand food shipped to the UK with some UK food. It asked which ones had most environmental impact. Results? Kiwi dairy products used half the energy of the UK ones. Kiwi lamb used a quarter of the energy of UK lamb. Kiwi apples were 10 percent more energy-efficient than UK ones.

Example 2

Each packet of green beans from Kenya in UK shops has a sticker with the image of a plane on it. This shows getting them to the UK cost emissions of carbon dioxide from jet fuel. However, Kenyans produce beans in an environmentally-friendly way. They use manual labour, cow manure fertiliser, and low-tech irrigation. Bean production gives jobs to many people. UK farmers plough bean fields with diesel tractors and spray with oil-based fertilisers. The expert verdict? Driving 6.5 miles to buy your shopping emits more carbon than flying a pack of Kenyan green beans to the UK.

ISBN: 978-0170210348

Example 3

British apples are harvested in September and October. Some are sold fresh. The rest are chill stored. The amount of energy used to keep them fresh exceeds the carbon cost of shipping apples from New Zealand. It is therefore better for the environment if British shoppers buy apples from New Zealand in July and August rather than British ones.

Example 4

You can buy dried or cooked tinned chickpeas. The carbon footprint of cooked chickpeas (heat needed to cook them before they are tinned) is higher than that of dried ones (just dry them in the sun). But you can't eat dried chickpeas. You have to cook them. Cooking with gas is more efficient and less likely to produce carbon emissions. Whatever method you use, you'll find the carbon emitted from cooking dried chickpeas exceeds that for the tinned chickpeas. That's because cooking small bits at home is inefficient compared with big industrial kitchens. If you used renewable energy to heat your home and kitchen, you would probably take away all the carbon impact of cooking them.

Even if food miles were an issue, New Zealand could be a world leader in sustainable food production. This is what it could do.

Keep the soil fertile (good for growing). Use natural ways to fertilise.

Treat workers fairly and make sure they can make a living from their work.

Stay away from factory farming which at present often has problems such as suffering animals and huge antibiotic use.

Recycle waste and make sure untreated waste does not get into water.

Be aware that food miles do not tell the full story.

Keep water clean and available.

Don't produce bottled water.

Use renewable energy.

Stay away from corporate farm owners who have less interest in and sense of belonging to communities than local farmers who want to look after their environment.

Produce heritage and heirloom foods. These are the thousands of different varieties people used to grow before Big Industry came along and concentrated on only a few varieties.

Make sure that the place the food comes from really is green and clean.

All around the world there are people working hard to make food production sustainable. One example is the Slow Food movement. Another is minilivestock farming.

ISBN: 978-0170210348

Case Study Slow Food

Slow Food is an international movement. It started in Italy when a man called Carlo got upset on seeing a McDonald's opening at the Spanish Steps, a famous historical part of Rome.

Slow Food supports:

- the survival of traditional crops and farming methods
- teaching people about local food and traditional tastes
- teaching people about the pleasures of cooking
- setting up seed banks to look after heirloom varieties
- organising celebrations of local food
- teaching people about the risks of fast food
- teaching people about the problems of commercial outfits such as factory farms
- teaching people about the risks of relying on just one or a few crops and animals
- preserving family farms
- lobbying (working to persuade) against government funding of genetic engineering (changing the genetic structure)
- lobbying against the use of pesticides
- teaching gardening skills to students and prisoners.

Case Study Eat Your Grubs Up

- In many areas such as Thailand, Vietnam, Cambodia, China, Africa, and Mexico people eat creepy-crawlies like grasshoppers, caterpillars, bees, wasps, ants, worms, cicadas, huhu grubs, walking sticks, desert locusts, Bogong moths, witchety grubs, termites, dragonflies.
- There's even a word for eating insects – entomophagy. It actually includes arachnids (mainly spiders) and myriapods (mainly centipedes).
- Why do people eat insects? They like the taste. Insects have stuff in them that is good for people. Crickets, for example, are high in calcium. Sago palm weevil grubs are full of unsaturated fat.

ISBN: 978-0170210348

- In some places, people get frightened at the thought of eating insects. Reality television such as *Fear Factor* shows this. Yet those people eat insect cousins like lobsters and shrimp. Locusts are even called sky prawn. And people eat insects without knowing. The red colouring in food often comes from the cochineal bug. Fruit and vegetables can contain small insects. Millers who make flour don't spend time picking out the beetles and weevils in the grain before they mill.
- Minilivestock farming is growing insects for food. It can be much more sustainable than traditional farming.
  - Insects aren't fussy eaters. They don't turn their noses up at cannibalism. They don't have to be fed crops and have grass grown for them. For example, if you ran a scorpion ranch in your home, as many people do in some countries, you could feed the scorpions with crickets and mealworms.
  - Insects don't have to use a lot of energy to stay warm. They are better at converting food they eat into body substance. For example, beef cattle convert 10 percent of their food into body substance. German cockroaches convert 44 percent.
  - Insects reproduce faster than animals.
  - Insects need less water than animals.
  - Insects are not associated with water pollution, over-use of water, soil erosion, pesticide use, and antibiotics as traditional farming is.
  - Insects produce much less greenhouse gases than farm animals.
  - Insects have friends such as the United Nations which wants to reduce the amount of meat people eat and has held meetings about eating insects instead of meat.
  - Insects are easy to harvest.
  - Insects take up little space.
  - Insects are nutritious (healthy to eat). How funny, say some food experts, that we spray crops with toxic chemicals to kill insects that are more nutritious than the crops they eat.
- Some people may never try eating insects because their religion bans it, or they may have allergic reactions to them. But how do you get others to? Perhaps smash the insects up so they don't look like insects and use them as flour, or put them into pies and patties.
- Feeding insects to farmed animals such as chicken and fish which eat them naturally, could be a start.

ISBN: 978-0170210348

Recently there has been discussion in New Zealand about how sustainable indoor factory farming is.

1 Write meanings for the following: locavore, food miles, corporate farm owner, fertile, heirloom food, lobbying, genetically engineered, minilivestock farming, entomophagy, arachnids, nutritious.

2 What are the two sides to the food miles issue?

3 Design a place mat to use at a meal featuring sustainable Kiwi food for overseas Ministers of Trade.

4 Explain the differences between fast food and slow food.

5 Prepare a chart about how how sustainable a food choice insects are. Here is an example of headings you could use.

INSECTS	TRADITIONAL ANIMALS	MY THOUGHTS

6 Describe how sustainable food production is a mix of environmental, social and economic features.

ISBN: 978-0170210348

Sustainable Gardening for Food

Permaculture is a method of sustainable gardening.

Permaculture

The word comes from permanent (lasting) culture (growing things).

Don't use pesticides.

Make a balance of plants, animals and insects.

Don't put crops into rows or beds. Mix them in with fruit trees and flowers so they are not in one place for pests to get established.

Use mulch such as straw or seaweed so there is no or little need to dig. Mulch means the soil does not crust and crack, weeds do not grow, and plant roots stay cool.

Reduce the need to irrigate. Xeriscaping, from the Greek word *xeros* that means dry, is about using mulch and compost as ground cover. Don't use rock as it will make the area hotter. If you have to water, do it deeply to develop deep roots. Never water during the day because you'll lose water to evaporation. Put in a rain sensor to shut off water when rain comes. Avoid lawns. If lawn is a must, choose grass that needs least water. Use a hand mower.

Design gardens that copy the way nature works. For example, chickens clean up most garden pests and produce manure which is good for the soil. They eat insects and worms from a worm farm, and kitchen scraps. They produce eggs with no antiobiotics or chemicals in them.

Put plant and vegetable wastes into compost bins to make compost that helps soil keep water in.

Rotate crops (grow them in different places each year) as different crops take different nutrients from the soil. Pests of one crop can get set up in an area and have big parties if crops are grown in the same place year after year.

Use renewable systems such as solar power and aquaculture.

Use vertical space such as training vines along walls.

ISBN: 978-0170210348

Case Study Vertical farming

Rice, wheat, corn and potatoes make up more than half the world's food supply. Much of it involves irrigated water, synthetic (manmade) fertilisers and pesticides, packaging, transporting, selling. Often the environmental costs are loss of topsoil, land erosion, land becoming desert from over-cultivation, use of fossil fuel, pollution of water and land.

By 2050 about 80 percent of Earth's population will live in urban areas.

Scientists suggest future cities will use vertical farming to grow food inside skyscrapers, which would then be called farmscrapers. All the technology needed for vertical farming exists. Scientists now have to work out how to put all the technologies together. This is how scientists think the system will work.

No fertiliser runoff pollutes rivers and oceans.

No forests are cut down to make plots to grow food.

It reduces the chance of infectious diseases destroying crops or small lifestock.

It uses greenhouse methods such as hydroponics.

Tanks house fish and other seafood.

Each floor features a variety of crops and even small livestock.

Big solar panels make energy.

Temperature can be controlled.

Methane gas is collected and turned into energy. Excess energy is sold to the local energy grid.

It saves land as one vertical farm hectare grows much more than one traditional farm hectare.

No herbicides, pesticides, or fertilisers are needed.

Crops grow all year.

Food is sheltered from droughts and floods.

Non-edible parts of plants are composted.

Glass walls let in light.

It is 30 to 40 stories tall.

Monitoring systems make sure energy and water go where they need to go.

Incinerators burn waste to make energy.

Irrigation systems send water exactly where it is needed.

Dew is collected through evaporation.

Excess water is collected and recycled.

Some sewage is cleaned by algae and plants and made potable (suitable to drink).

Some sewage is treated by filters and turned into grey water for irrigation.

Land and water are not polluted by crop growing.

There are no pests.

Land can return to forest and natural ecosystems such as wetland.

ISBN: 978-0170210348

1 You are going into business as a supplier of mulch and/or compost. Make up a name for your business. Then make up a full page advertisement for your business to go in your local newspaper.

2 Make a small circle in the middle of a page. In it put 'Vertical Farming'. Make a big circle around it. Divide it into segments. In each segment put a result of vertical farming.

3 Design a set of five postage stamps that shows New Zealand as a world leader in sustainable gardening for food.

4 You have scored a job as a guide to school parties in either a permaculture garden or a vertical farm. Make a list of things you will show the students.

5 In your group make a big sheet that will show how people's ideas about food farming are changing. Use two shapes to show this.

A circle holds the idea people used to have.

A square holds the idea people now have or may have in the future.

The rest of the design is up to you.

6 Use the internet to find out more about sustainable gardening. Decide on five key questions you want answers for. Next decide on five key words or terms you will search under.

ISBN: 978-0170210348

Softly-Treading Tourists

Sustainable tourism aims to have eco-friendly tourists who are kind to the environment and kind to the local people.

Actions that make you an eco-friendly tourist.

- Respect the local culture (eg. don't make a show of gagging and spitting if a local gives you something to eat and then tells you it's a cockroach).
- Buy local goods (eg. souvenirs made locally rather than ones made overseas).
- Do business with organisations that are environmentally-friendly (eg. organisations that have the Green Globe logo which is the international brand for sustainable tourism).
- Take photographs with care (eg. ask permission first).
- Bring reusable containers so you don't leave waste (eg. bring a reusable water bottle and maybe purification tablets to reduce waste from plastic disposable bottles).
- Don't leave toxic waste (eg. pack rechargeable batteries as batteries are toxic and many places do not have proper disposal facilities).
- Protect local systems (eg. use local transport).
- Stay at locally-owned places (eg. try a local B&B rather than a chain hotel).
- Support fair trade (eg. buy goods directly from people who make them).

ISBN: 978-0170210348

ECO-FRIENDLY TOURISM

Having low impact so future generations can have the same experience as you.

Visiting places that humans haven't touched.

Wilderness adventures.

Volunteering to help locals on a project.

Learning new ways to live on the planet.

Gaining respect and love for the environment.

Avoiding luxury.

Examples in New Zealand

- gannet colony at Cape Kidnappers
- muttonbirds at Stewart Island
- yellow-eyed penguins at Otago
- kayaking in Fiordland
- diving to the Rainbow Warrior wreck in Matauri Bay
- walking in the Waipoua Forest.

Using and learning from expert guides.

Taking responsibility for the environment.

Seeing living organisms in their natural environment.

Fastest-growing market in the tourism industry.

An example in Australia is a Wildlife Walkabout. You learn about the history and culture of local Aboriginal people. Guides take you to see kangaroos, koalas, parrots, lyrebirds, gliding possums, wallabies. You travel into eucalypt forest to see huge goanna lizards. You visit rainforest, untouched beaches and rivers, and a refuge for injured and orphaned wildlife.

The latest sustainable holiday attraction is the staycation. (Tourists go on vacation, which is where the term comes from.) Firstly, you plan your staycation and get excited about it. On the day of leaving you go to a place nearby, such as a restaurant where you have a meal. Then you go back home and begin your staycation. There you ignore all your usual routines, chill out and act as a tourist. The idea is to eat what and when you like, go to bed when you like, turn off all phones and computers.

Case Study White Island (Whakaari) Tour

- This tour has won Green Globe Certification which shows it is into sustainability.
- It explores the inner crater of New Zealand's only active marine volcano, 48 km off Whakatane in the Bay of Plenty.
- Boats carry biodegradable motion sickness bags.
- It has resisted pressure to build toilets and other facilities on the island.
- It has a rule that nothing gets taken off the island and nothing is left on the island.
- Boarding passes for the boat are reusable separator discs which used to be part of farm milking machines.
- The skipper obeys all the regulations about the boat having a low impact on mammals it meets such as whales and dolphins.
- Boats are anchored in ways that don't damage the sea floor.
- Food scraps go into a pig bucket for a local farmer to feed his pigs.
- It follows a single trail on the island for people to walk on.
- Boats save as much diesel as possible.

ISBN: 978-0170210348

- It does not allow overnight stays on the island.
- Anyone who stays at a tour company motel is encouraged to recycle waste, request less laundering, use drying racks, report leaks and turn off power. Most cleaning products come from Ecolab, whose products are 'environmentally preferable'. Staff know to put liquid waste down sewer drains, rather than stormwater drains. Shower glass has a protective cover so that it needs cleaning only with a dry towel. The tour company employs locals. It is trying to increase water and energy efficiency. It has appointed an environmental champion to make sure all environmental standards are kept up.

- It is searching for an alternative to its polystyrene cups to give passengers soup and hot drinks. These are used instead of paper cups because a paper cup consumes more than twice its weight in wood, 15 times more chemicals, 6 times as much steam, 13 times more electricity, 30 percent more cooling water, 170 times more process water. Food Safety laws stop the use of crockery cups because that needs things like a steriliser on board.

Case Study EarthCheck

Please Respect The Wildlife

Sometimes They Demand A Little Space And Privacy

- Thank You

EarthCheck

You've Chosen A Hydro-Powered Holiday

Thank You For Helping Us Conserve Energy And Reduce Emissions

EarthCheck

EarthCheck is a global programme that is recognised as the best in the world for giving certificates for sustainability in the travel and tourism industry. It measures things such as consumption of energy and water, waste production and community efforts to be sustainable.

Kaikoura District Council has taken up the EarthCheck Sustainable Communities Programme. Kaikoura's nature and activities such as whale watching attract nearly a million visitors each year. Residents want to protect that nature and make sure it's there for future generations. Here are examples of how they are doing that.

- They were the first in New Zealand to employ an environment officer.
- They are working to be a Zero Waste to Landfill Zone by 2015. They stopped

ISBN: 978-0170210348

kerbside rubbish collections to encourage recycling and offer a free weekly recycling pick-up service. A recycling depot works on electrical goods and wood for reuse. Kitchen scraps go into a community composting system.

- Each year a Trash-to-Fashion show celebrates how local artists use everything from fishing buoys to toothbrushes to make clothes, accessories and gifts.
- Biodiversity schemes include rates relief for owners who set up protected areas on their land.
- Trees for Travellers works by tourists buying native trees which are planted in the area. The tourist get GPS coordinates for trees so they can come back to visit them.
- Council staff have paper recycling bins but no rubbish bins at work areas; they print on both sides of paper, are issued with warm vests to reduce heating and teleconference to avoid long-haul travel to business meetings.
- Cleaning staff use biodegradable chemicals. Restrooms have cloth towels instead of paper towels and air dryers.
- A Warm Up Kaikoura project insulates homes.
- Waterways are improved by planting, fencing, bridging.
- Overfishing is reduced by a rahui (ban) on a section of sea for two years. Then that section opens and another closes.

1 Make a chart to show the advantages and disadvantages of a staycation.

2 List ways you can be a softly-treading tourist.

3 You own a tourist business. Decide what sort. Prepare an advertisement for an environmental champion or officer for your business.

4 Describe how Kaikoura is trying to improve its sustainability.

5 As an environmentally-friendly tourist give five items you would include in your suitcase and five items you would not. Then describe how, as a tourist, you have rights and responsibilities.

6 Find at least four eco-tours that you would like to do anywhere in the world.

ISBN: 978-0170210348

Sustainable Buildings Aren't Sick

Sick buildings are ones with ailments such as mould or leaky air-conditioning units. People who work or live in them get sick too. Sustainable buildings don't get sick. This does not mean they have to be caves or tree huts. You can still have your home comforts in a sustainable building. Banks are starting to offer 'green mortgages'. These let you borrow extra money to be more enery-efficient. Banks think these homes will keep their value over time better than non-sustainable homes.

In developing countries many people do live in zero-energy buildings such as yurts and caves. They are off-the-grid which means they are not connected to an off-site energy provider. Such zero-energy buildings are becoming more popular in the developed world. One design is called the Earthship.

An Earthship can be built in any climate just about anywhere. It's a sustainable house made of natural and recycled materials. Outside walls are stacked old car tyres with clay packed into them. Non-load-bearing inside walls are mortar with things like bottles and aluminium cans for decoration. The front wall is mostly glass sheets. It faces towards the equator to get the most sun. The sustainable timber trussed roof is covered with material that captures potable water. This drains into containers. Water flows through a system that filters it and pumps it into a tank. Used water goes to inside planters in which vegetables grow. Water that drains from planters is pumped into the outdoor compost toilet. Solar hot water panels on the roof give hot water. Electrical power comes from sustainable generation on site such as solar, wind, and mini-hydro.

ISBN: 978-0170210348

A green roof has a waterproofing membrane and vegetation planted on top of it. It takes in rainwater and reduces stormwater runoff. This helps protect against flood impacts and streambank erosion. It makes a habitat for wildlife such as birds, bees, and butterflies. It helps lower air temperatures. It filters pollutants and carbon dioxide out of air, and pollutants and heavy metals out of rainwater.

Case Study: Conservation House (Whare Kaupapa Atawhai)

This is the Department of Conservation's head office in Wellington.

It is designed for sustainability. It is recycled because it used to be a cinema complex. It has eco-friendly fittings such as wooden doors. Its framework and panelling come from sustainable forests. There is a lot of natural light, eco bulbs and lighting sensors. A large atrium runs through the whole building, letting in more natural light, and airflow.

Rainwater is collected for toilets, cleaning, and gardens. The roof is green. A cafe in native gardens lets staff go outdoors without leaving the building.

A system of active chilled-beams controls temperature. Warm air rises past water-carrying coils in ceiling beams. As cool air comes down, rising warm air replaces it and creates an airflow cycle. In winter heated water running through the coils warms the surrounding air. Heat pumps capture waste heat for water-heating.

It aims to have as little as possible impact on the environment. Soundproof eco-panels in the meeting rooms are made from recycled milk bottle tops. Staff are encouraged to use stairs rather than lifts. They have no personal rubbish bins. Instead there are recycling stations and organic waste bins. Windows have double cladding. The floor on the stairwell is from recycled car tyres. There is a big bike rack. A wind turbine on the roof powers the roof lights.

ISBN: 978-0170210348

1 Decide how important your house and place of work, such as school and any after-school job, is in your life. Give reasons for your decision. Now give advantages, and any disadvantages, of living and working in sustainable buildings.

2 You want to apply to your bank-manager for a loan to build a house. Make a list of sustainable things you are putting in your house so you can ask for a green mortgage.

3 Explain how the following can be associated with sustainable buildings: yurt, bank, car tyre, waterproof membrane, double cladding, atrium, chilled-beams.

4 Create a design for a non-load-bearing wall using mortar and recycled materials.

5 Collect six photos of sustainable buildings anywhere in the world. Rate them in order of the ones you would like to visit most.

6 Find out about other types of sustainable houses such as straw bale ones.

ISBN: 978-0170210348

Shaking up Cities

Cities house 50 percent of the world's population. They consume 75 percent of the world's resources. They emit 75 percent of the world's greenhouse gases.

SmartGrowth is an idea to make cities and the areas around them more sustainable. It says growth should be concentrated in the middle of the city or town. People then find it easier to walk, bike or use public transport to get to work or to entertainment. That means less pollution and less fuel use. Vegetation called green belts stops urban sprawl, makes a habitat for wildlife, and helps clean air and water.

Case Study Western Bay of Plenty in New Zealand

Every week in the Bay:

- 100 people arrive to live
- 52 people leave
- 32 new houses are built
- 54 more vehicles go on the road
- 45 new jobs are created.

By 2051 the number of traditional families will double, single and two-person households will treble, people aged 80 or older will rise six-fold.

ISBN: 978-0170210348

THE CAUSE

By 2000 people in the community were worried. How sustainable was this population growth?

THE RESULT

Environment BOP, Tauranga City Council, Western Bay of Plenty District Council, tangata whenua (local Maori) consulted their communities. All decided on a SmartGrowth plan to get sustainable development. It involves many people such as environmental planners, strategic planners, policy developers, water and transport engineers, communications advisors, Maori advisors.

THE PLAN

- It is a 50 year plan and describes how the area will look in 2051.
- There will be improvements to water resources, fisheries, indigenous plants and animals, systems dealing with waste.
- Open Days will have been held for people to come and talk about the plan.
- The number of dwellings per hectare will go from 10 to 15 or more.

Another sustainability idea is the eco-city which is a sustainable city. Scientists and planners know the things that will turn cities into sustainable areas:

- more renewable energy and less fossil fuels
- more natural products and less synthetic chemicals
- people seeing the environment and its resources as something to respect rather than trash
- designs that consider environmental impact
- no pollution
- new farming systems such as vertical farms and agricultural plots within the city
- green belts
- windows that open rather than air conditioning
- green roofs.

Waitakere was New Zealand's first eco-city. Its vision is more public transport, more cycling, more walking, less energy use, renewable energy, resources used more wisely and producing less waste. It runs workshops on Sustainable Living. Its Sustainable Living Centre has information, displays, workshops, an eco library, and advice on how to become more sustainable at home, at work, at school, in a community. It has organic gardens divided into zones such as Compost Zone, Food Forest, Maori Medicinal Zone. It has working compost and worm bins. In 2010 Waitakere became part of the Auckland Supercity.

ISBN: 978-0170210348

Case Study World's Top Eco-City

Maybe this year your city or the city nearest to you will win this title. Judges look at water availability, water potability, waste removal, sewage, air pollution, and traffic congestion. They say a high-ranking eco-city uses more renewable energy than non-renewable energy and generates the least possible pollution. In 2010 Calgary in Canada was the top eco-city in the world. Honolulu was second, and Ottawa and Helsinki tied for third. Calgary had better waste removal, sewage systems and drinking water. It had less pollution and traffic jams. Calgary has just opened a new $430-million Wastewater Treatment Centre. It will meet the needs of its growing population for the next 10 years. However, city leaders say Calgary cannot think it's totally sustainable. Some days, for example, it has brown smog.

1 Make your own copy of the Sustainable Development diagram. Choose the colours and/or patterns you use for it to stand for something. Be ready to explain your diagram to the class.

Sustainable development

Economy

Society

Ecology

2 How sustainable are global cities at present? Give some proof to back up your answer.

3 In your group make up a short mime or play about encouraging more sustainable transport in a city.

4 Make up a set of headings that judges could use if there was a competition for New Zealand's Top Eco-School.

5 Name the communities involved in the Western Bay SmartGrowth project. Explain the causes of the project and the hoped-for results.

6 Find out what plans your Council has for the future sustainability of your area.

ISBN: 978-0170210348

Mining Doesn't Have to be End of Story

Mining has big impacts on the environment. To get at the ground, miners have to clear or burn trees and vegetation, and drain waterways. They may need to build roads and airstrips. In come heavy machines. Mining processes use chemicals such as cyanide and mercury which often end up in rivers and oceans. Greenhouse gases go into the air. Mining pits fill with water that becomes a breeding ground for mosquitoes and other insects. When the miners move on they can leave behind a toxic environment.

Mining is an emotional issue – people have strong opinions for or against. Recently a huge public outcry forced the New Zealand Government to back down on its proposal to mine in National Parks. One possible future issue is mining on the moon. Countries have already talked of mining helium-3 from the moon to use as fuel in nuclear reactors.

ISBN: 978-0170210348

Case Study Sierra Leone

Sierra Leone is a very poor country in Africa. Its diamond mining is associated with rebels (people fighting the government) chopping off arms and legs of miners, boy rebel soldiers and blood diamonds - diamonds mined in areas controlled by rebels and sold to get money for the fighting.

Most of its diamonds are close to the surface. Miners have just needed a shovel, bucket and sieve. Yet even this low technology has left an environmental mess. Pools of mosquito-infested water, tens of thousands of mining pits, huge buried tree roots from where trees were chopped down, land where nothing grows.

The Foundation for Environmental Security and Sustainability (FESS) works to improve environmental sustainability around the world. It is about reclaiming land after mining and developing it for sustainable agriculture. FESS set to work in Sierra Leone. It made demonstration sites as models of how damaged land could be reclaimed. People can use mining skills to make sustainable land.

The whole community gets involved – chiefs, men, women, young children. They get a weekly payment. The women cook lunch on site for the workers and bring in water for them to drink. The community becomes responsible for security on site such as looking after the tools. The project gives jobs for youths, many of whom were bored and doing drugs. The community sees the link between themselves and the future.

Firstly the community agrees that diggers are no longer finding diamonds. Then they fill in all the mining pits in the area. As soon as they've finished, they plant crops such as rice, oil palm seedlings, and vegetables. This is so other people don't start re-mining the reclaimed land.

ISBN: 978-0170210348

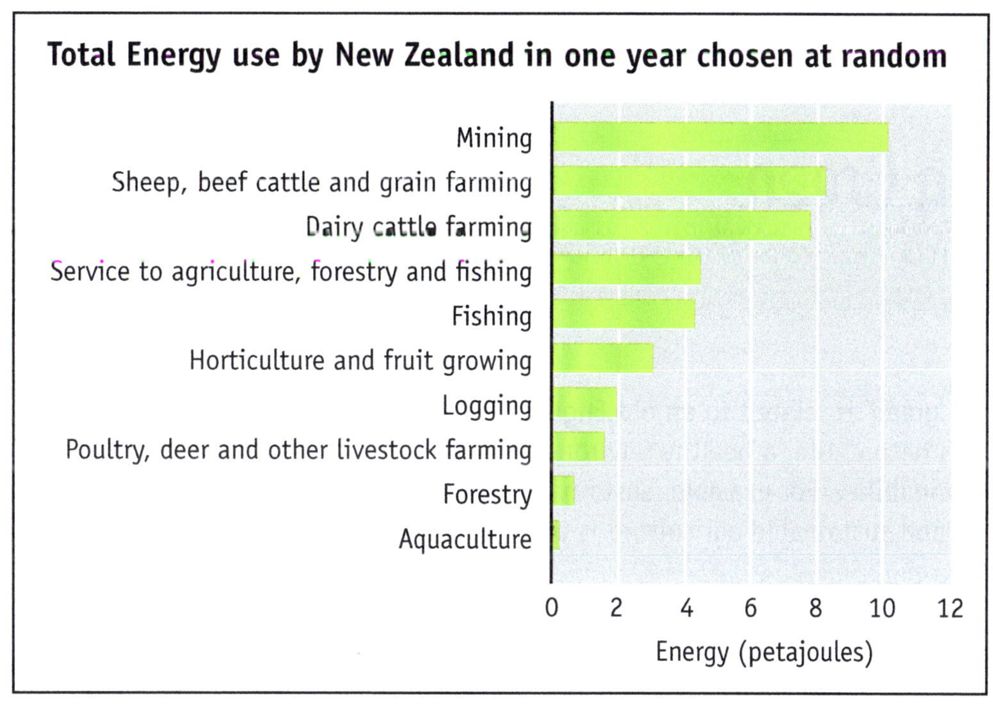

Mining is a heavy consumer of energy. In New Zealand a lot of the energy is used in running machinery in off-road environments.

1 Explain why mining is called an extractive industry.

2 Give the approximate figures for energy use by each of the industries on the graph.

3 Helium-3 is almost unavailable on Earth. Should countries be allowed to mine on the moon? Interview your partner to get his or her ideas. Then work out the class percentage of Yes and No votes.

4 Give reasons and examples of why mining can be an emotional issue.

5 Describe the work that FESS does. Give an example.

6 Find out what mining companies are doing to improve sustainability.

ISBN: 978-0170210348

Going Green

The word 'green' is related to an old English word meaning to grow. Green represents nature, life, a healthy environment. The term is often used with sustainability. For example, sustainable building is often called green building, and sustainable agriculture is often called green agriculture.

Green Cross International was founded by a former Russian leader to help get a sustainable future for all. It aims to prevent fighting over scarce natural resources.

Governments picked up the idea and founded their own Green Cross organisations.

Eg. Green Cross Australia believes we must help each other to help our planet. It wants to reconnect people and environment.

GCA is involved with Extreme Weather Heroes. They are young people who know about climate change, and risks of an upset weather future. They are volunteers, not just for frontline action such as fighting bushfires, but also geeks and admin people.

They try to inspire those hit by weather disasters to rebuild green. The typical Aussie home generates 14 tonnes of greenhouse gas each year. Even something as simple as replacing light bulbs with more energy-efficient ones is helpful.

Green Cabs is an environmentally-friendly taxi company. It runs a fleet of hybrid cabs in Auckland, Wellington and Christchurch. It plants trees to offset the emissions it produces. It does this in advance so when you get into one of its taxis you know you're being kind to the environment.

ISBN: 978-0170210348

Scientists are inventing green products that use materials such as mushrooms, rice husks, and seaweed to make biodegradable versions of disposable items. Rice-and-mushroom packaging, for example, uses a lot less of the energy needed to make the same amount of styrofoam and is biodegradable. You fill a mould with agricultural waste like rice husks, and add mycelia (fungi like mushroom roots). Within two weeks the roots have grown to make a thick, lightweight network stronger than styrofoam. Scientists say people will be able to use this technology to grow their own packing material almost anywhere. It will be like baking bread.

An example of a green technology is the Swedish company eGain. It uses forecasting technology to predict the future weather's impact on a building. By changing the heat based on the weather forecast its system prevents unnecessary use of heat, which reduces energy consumption and greenhouse gas emissions.

NASA (The US National Aeronautics and Space Administration) studied which plants were best to grow indoors to keep the air fresh and filter common dangerous air pollutants such as formaldehyde and ammonia which can cause asthma, allergies, and sick buildings. It put house-plants in sealed chambers and exposed them to hundreds of chemicals. Among the top green plants were Spider Plant (above), Peace Lily, Rubber Plant, Bamboo Palm, Snake Plant, Weeping Fig.

Examples of green gifts	Examples of green shopping
• re-gifting (not to the person who gave it to you) • second-hand goods such as china • a third party gift (telling the person you donated money in their name to a charity) • giving a gift of your time such as gardening • making your own gift such as a cake • using material such as wool instead of cellotape, and newspaper as wrapping.	• using canvas bags • buying in bulk and storing it • not buying bottled water • buying resuable and long-lasting items such as non-disposable cameras and cloth instead of paper products • buying clothes made from renewable materials such as organic cotton and natural dyes • buying locally • gift vouchers.

ISBN: 978-0170210348

Case Study Greenwashing

Greenwashing is spin. It's putting a picture of a mountain stream on a bottle of toxic chemicals, or zoos trying to get more visitors by claiming they are leaders in wildlife conservation when hardly any animals born in zoos are introduced to the wild. It's telling customers you'll no longer include paper manuals with video games because you want to protect trees when really you want to save the company a lot of money.

It's telling shoppers that your brand of toilet paper and tissues is certified as being environmentally sustainable yet you have no certificate and the company that supplied the pulp for your products got the timber from an Indonesian rainforest.

It's saying your product is free of a particular chemical, yet that chemical was banned 25 years ago anyway.

Sustainability communications firm Futerra has a Greenwash Guide. It says there are 10 signs to look for. They are:

- fluffy language (words or terms with no clear meaning)
- green products versus dirty company (eg. efficient light bulbs made in a factory which pollutes rivers)
- green images that suggest a green impact that isn't real (eg. flowers in exhaust pipes)
- unimportant claims (emphasising one tiny green feature when everything else is not green)
- best in class (saying you are greener than the rest)
- just not believable (eg.how safe would eco-friendly cigarettes be?)
- gobbledegook (language and information that only a scientist can understand)
- imaginary friends (a label that looks like a third party recommends the product but there is no such third party)
- having no proof for claiming to be green (eg. no certificate)
- lying (making up data and claims).

ISBN: 978-0170210348

Case Study New Zealand Green Ribbon Awards

The Ministry for the Environment runs Green Ribbon Awards each year to recognise outstanding efforts in looking after the environment. Here are winners for 2010.

Supreme Winner: Kaharoa Kokako Trust. It helps kokako recovery. Volunteers use safe, ground-based pest control to attack kokako enemies.	***Category:*** Protecting our Biodiversity ***Winner:*** Banks Peninsula Conservation Trust. It works with landowners to look after and restore biodiversity on private land.	***Category:*** Reducing our Carbon Emissions. ***Winner:*** Landcare Research - carboNZero Programme. It specialises in sustainable management of land and biodiversity.
Category: Caring for our Water. ***Winner:*** Waitakere City Council - Project Twin Streams. Local groups such as schools are helping to restore 56kms of stream bank.	***Category:*** Managing our Waste. ***Winner:*** Envirofert Limited. It operates one of the biggest composting sites in New Zealand.	***Category:*** Improving our Air Quality. ***Winner:*** Sleepyhead Manufacturing Company Limited. New technology makes bedding products more environmentally-friendly.
Category: Community Action for the Environment: Young People. ***Winner:*** Hukanui Primary School - Eco-classroom. It designed, funded and built 'The Living Room'.	***Category:*** Community Action for the Environment: Volunteers and Not-for-Profit organisations. ***Winner:*** Environmental Education for Resource Sustainability Trust. Its Paper4trees encourages recycling in schools.	***Category:*** Protecting our Coasts and Oceans. ***Winner:*** Sustainable Coastlines Incorporated. It is run by young people whose mission is to sustain and protect our coastlines.
Category: Environmentally Responsible Large Organisations. ***Winner:*** Resene Paints Limited. It develops environmentally-sustainable surface coatings.	***Category:*** Environment in the Media. ***Winner:*** Splashroom Limited. Its vision is to be NZ's top producer of films to help solve social and environmental challenges.	***Category:*** Public Sector Stepping Up. ***Winner:*** Palmerston North City Council - Awapuni Sustainable Development Centre. This site used to be a landfill; now Council works to reduce the waste.
Category: Small Businesses Making a Difference. ***Winner:*** The New Zealand Wine Company Limited. It became the first carbon-neutral certified-winery in the world.		

ISBN: 978-0170210348

1 Choose any colour except green. Try to justify (give reasons for) using your colour instead of green to represent sustainability. Share with the class.

2 Explain the connection to green that the following have: re-gifting, third-party gift, mycelia, Spider Plant, canvas bag.

3 People used to roll their eyes at Greenies. Is this attitude changing? Give proof for your answer.

4 Make a list of the companies and organisations mentioned in this unit. Decide which one you would prefer to work for. Give reasons for your answer.

5 A product has a bright green sticker on it saying it is CFC-free. It doesn't mention that CFCs were banned many years ago. What is that an example of? What are other signs of this?

6 In 2010 a campaign began to have the United Nation's International Criminal Court recognise ecocide – the thoughtless or deliberate destruction of the environment – as a crime against peace. Find out about the campaign and also find out if countries have laws about and punishments for misleading environmental claims.

ISBN: 978-0170210348

China is a Test Case for Sustainability

In the last 50 years, China has become a giant.

It used to consume a reasonable amount of Earth's resources. Now it consumes a lot. Its ecological footprint has quadrupled.

It now demands more from Earth than any country except the US. If China follows the consumption habits of the US, it will soon demand the available resources of the whole planet.

Therefore China is going to be a big player in the world's battle for sustainability.

If it can get its act together, it could lead the way to a future where people and the environment are best mates. China is like a test case for the rest of the world. If China can do it, so can everyone else.

The cause

China began to grow by putting vast numbers of people to work in factories. They got paid little but produced much. Their cheap goods went overseas. Other countries kept asking for more. At that time sustainability was not a big topic.

ISBN: 978-0170210348

The result

For many years China opened a new coal-fired power plant every week. Some rivers turned black. Some land got poisoned. Some forests got chopped down. Pollution filled some of its sky and caused a decrease of rain in the mountains. China sent a brown smog over other parts of Asia and even further. In some parts of China pollution became a vicious circle. Polluting industries made some people rich. Rich people bought more cars. More cars emitted more pollution. KFC became the biggest restaurant chain in China. Some Chinese got obese.

The reaction

China claimed it should still get money donated to it from rich countries who give aid to poor countries. It became one of the biggest borrowers from the World Bank. When other countries ask China to agree to reductions in carbon emissions, China says it can't because it's still a developing country.

The Chinese Government hands out environmental awards and names environmental model villages and cities. An example of how accurate these can be is China's freshwater lake, Lake Tai. Algae bloomed there. It blocked water supplies to millions of people in summer. Yet many polluting factories in the area had passed environmental inspections and authorities said they had shut down all polluting factories along the lake shore. An environmentalist who spent years collecting water samples from the lake and warning about rising pollution was sentenced to prison on charges that he blackmailed polluters.

In one village villagers objected to the building of a polluting smelting factory. Authorities forced them to move to make room for it. In another village, people got no response from authorities when they kept asking them to close a polluting waste dump. They tried to block the road to the dump. Armed police sent them away.

Not long ago Linfen in China won the title of most polluted place on Earth. Experts compared the effects of its pollution to that of a nuclear power plant meltdown. Well over four million people lived in Linfen. So did thousands of coal mines, steel factories, coal-burning power plants. When the Government ordered one plant to shut down, another one opened up illegally. People wore masks. Some days they couldn't see their hands in front of their faces. Their vehicles needed headlights on during the day. Any snow turned black. The Fen River, source of drinking water for millions, was an open sewer. Some waterways were so badly polluted that even industries couldn't use the water. People had a high rate of arsenic poisoning and breathing problems. Efforts are being made to clean up Linfen, but there is still a long way to go.

ISBN: 978-0170210348

However, there's also good action in China. Awareness of sustainable development and how to get it is spreading in Government and society. Some experts say that China is trying so hard that as it travels forward down the road to sustainability, the US travels backwards in comparison. One example is Suzhou, a very old city in China. It's the second biggest industrial city, after Shanghai. The mayor says its people have a responsibility to develop the place sustainably. They should not use up resources that belong to the next generations. People are starting to pick up litter. Roofs have solar panels. Street sellers advertise solar heaters. Signs in hotels and restaurants say 'Water is the people's resource'. There is recycling of metal, newspapers, bottles, styrofoam. One company worked with schools on activities such as collecting used batteries for recycling and cleaning up waterways and canals. Suzhou Industrial Park has developed an Eco Action Plan. It has an Eco Science Hub to demonstrate low-carbon development.

Case Study The Sino-Singapore Tianjin Eco-City

It is now estimated that in just five years, over 50 percent of China's people will live in cities. Singapore is a world leader in sustainable technologies. Its government has an agreement with the Chinese Government to develop an environmentally-friendly city in China. It aims to show how both countries are determined to get sustainable development. The city will be a model for sustainable development for other cities in China.

ISBN: 978-0170210348

Tianjin Eco-City

- About 350,000 people live there.
- It is located 40 km from Tianjin city centre and 150 km from Beijing.
- It is on land that was one-third saltpan, one-third deserted beach, and one-third water, including a 270ha wastewater pond.
- A lot of its water comes from non-traditional sources such as desalinated water.
- The wetlands and biodiversity that are already there are looked after.
- There are green spaces.
- There is reduction, reuse and recycling of waste.
- People on lower incomes get help so they have houses near to those on higher incomes, and socialise with them.
- Public transport is light-rail, trams and buses.
- There are no barriers to stop people walking, biking, and using wheelchairs.
- It respects local heritage such as the two villages already there, and the Ji Canal which is 1,000 years-old.

1 Draw small labelled sketches to show you know the meaning of these: obese, smog, vicious circle, styrofoam, blackmail, saltpan, desalinated water.

2 You could go to China for work experience. For example, students recently worked with locals on a project in South China to evaluate sustainable tourism development there. If you were going on a work experience, what kind of project would you like to work on? Explain your choice.

3 Give evidence that coal provides much of China's energy.

4 Explain why life expectancy in Linfen is lower than the Chinese average.

5 Name some things that show China is becoming environmentally-aware and working towards sustainability.

6 Find out more about China and its environment.

ISBN: 978-0170210348

The Earth Charter

A global network of people, groups and organisations who were worried about Earth's sustainability got together and wrote The Earth Charter. Here are a few of its ideas.

Protect and restore Earth's ecosystems and biodiversity.

Make sure consumers can identify products that meet the highest environmental standards.

Push for the open exchange and use of new knowledge.

When knowledge is limited, go cautiously.

Push for environmentally-friendly technologies.

Make those responsible for hurting the environment pay.

Make sustainable development plans.

Make sure that ecosystems can deal with waste.

Get rid of and stop the introduction of non-native or genetically modified organisms harmful to native species and the environment.

Manage the mining and use of non-renewable resources such as minerals and fossil fuels in ways that cause no environmental damage.

Avoid military activities damaging to the environment.

Look after Earth's powers of regeneration.

Make those who argue that a proposed activity will not damage sustainability prove it.

Look after natural heritage.

Manage the use of renewable resources such as water, soil, forest products, and marine life in sustainable ways.

Reduce, reuse, and recycle materials used in production and consumption.

Use more renewable energy sources such as solar and wind.

Use energy efficiently.

Support international cooperation on sustainability, with special attention to the needs of developing nations.

Advance the study of ecological sustainability.

Allow no build-up of radioactive, toxic, or other dangerous substances.

Prevent pollution of any part of the environment.

Help the recovery of endangered species.

ISBN: 978-0170210348

Case Study Wondai

Wondai is a small rural town in Australia's Queensland. An ongoing drought makes water an extra-important resource. Students at the local school are from six to 15 years old. Their motto is 'Deeds not Words'.

Students studied a unit linked to the Earth Charter. They thought about three questions.

- How will a more sustainable world help me?
- How can we use the resources in our environment and maintain biodiversity?
- How can our water resources be used more sustainably?

Students brainstormed actions they could do as a class. For example, they organised a 'walk to school' day to fight greenhouse gas emissions. They made and displayed banners and posters. They ran an Earth Charter stall at the school fete. They organised Earth Charter wristbands for staff, students and the community to wear.

Students wrote this Earth Charter song for their school.

This is our future, this is our home
We need to care for this great biome
Respect each other, our planet too
Enough for us, enough for you
This is the Earth Charter of Wondai School
It's how we want to live and it's pretty cool
We have enough for us, plenty more to share
We want the whole wide world to know we care
This is our vision, a sky of blue
Clean air to breathe, for us for you
Healthy forests, water that's clean
Sparkling oceans, paddocks of green
Care for each other, respect all life
We want world peace, not war and strife
No threatened species, habitats for all
We need to plant trees, not make them fall
We are all equal, we all have needs
Care for our Earth, we'll take the lead
Respect our planet, each other too
Enough for us, enough for you.

Students did an electricity audit of the school. They made a plan to reduce energy consumption.They put a scrap paper box in each classroom. Recycled paper made things like notepads and carrybags. Another activity was to form four teams - Stormwater, Litter, Water Use, Energy. Examples of Stormwater actions were designing and painting 'Clean Seas for Me' signs next to stormwater drains to enourage students not to litter, and getting rid of leaves in school drains.

ISBN: 978-0170210348

1 Read the Preamble of the Earth Charter and answer the questions that follow.

> Preamble
> We stand at a critical moment in Earth's history, a time when humanity must choose its future. As the world becomes increasingly interdependent and fragile, the future at once holds great peril and great promise. To move forward we must recognize that in the midst of a magnificent diversity of cultures and life forms we are one human family and one Earth community with a common destiny. We must join together to bring forth a sustainable global society founded on respect for nature, universal human rights, economic justice, and a culture of peace. Towards this end, it is imperative that we, the peoples of Earth, declare our responsibility to one another, to the greater community of life, and to future generations.

- **a** Would the Preamble go at the beginning or at the end of the Earth Charter, and why?
- **b** In just one sentence, give the message of this Preamble.
- **c** Why is mankind at a critical moment?
- **d** What four things should form the basis of a sustainable global society?

2 Create, or find, a graphic that could go with the Preamble.

3 Explain the difference between an audit and a plan, a motto and a song, a biome and a planet, a banner and a fete.

4 Find as much proof as you can in this unit that the people who wrote the Earth Charter were looking out for your future.

5 Decide on actions your class could do to help the school environment.

6 Design an Earth Charter wristband.

ISBN: 978-0170210348

Sustainability Will Save the Planet

Here are some examples of comments about sustainability from ordinary people and famous people.

There are no passengers on spaceship Earth. We are all crew.
– Marshall McLuhan

Anything else you're interested in is not going to happen if you can't breathe the air and drink the water. Don't sit this one out. Do something.
– Carl Sagan

I can't understand why people are frightened of new ideas. I'm frightened of the old ones.
– John Cage

If you want to see an endangered species, get up and look in the mirror.
– John Young

When we tug at a single thing in nature, we find it attached to the rest of the world.
– John Muir

Treat the Earth well. It was not given to you by your parents. It was loaned to you by your children.
– Kenyan proverb

We are living on this planet as if we had another one to go to.
– Terri Swearingen

Sustainability is here to stay or we may not be.
– Niall Fitzgerald

I think the environment should be put in the category of our national security. Defense of our resources is just as important as defense abroad. Otherwise what is there to defend?
– Robert Redford

We could get off the couch and become an ACTIVIST! Go on – become the change you want to see in the world.
– Annie Lennox

By employing the intelligence of natural systems we can create industry, buildings, even regional plans that see nature and commerce not as mutually exclusive but mutually coexisting.
– Brad Pitt

Our planet's alarm is going off, and it is time to wake up and take action!
– Leonardo DiCaprio

ISBN: 978-0170210348

My hope is that we come to see consumption as slightly naff, something you only do when you have to.
– Chris Goodall

There would be very little point in my exhausting myself and other conservationists themselves in trying to protect animals and habitats if we weren't at the same time raising young people to be better stewards.
– Jane Goodall

Don't throw anything away. There is no 'away'.
– Royal Dutch Shell

We are each as much a part of the environment as a tree is. The ecosphere is an interconnected web (like the Internet).
– Matt Howes

Climate is an angry beast and we are poking at it with sticks.
– Wallace Broecker

Tell me, I'll forget. Show me, I may remember. But involve me and I'll understand.
– Chinese proverb

What good is a house, if you haven't got a decent planet to put it on?
– Henry David Thoreau

Business and the environment: Who Cares, Wins.
– Unknown

When written in Chinese the word crisis is composed of two characters. One represents danger and the other represents opportunity.
– John F. Kennedy

Sustainable development: Growth without cheating our children.
– Unknown.

This is the new politics. Personal responsibility. Not leaving it to others. I am my planet's keeper.
– Hilary Benn

The whole world is our dining room, but be careful: it is also our garbage can.
– Ashleigh Brilliant

My interest is in the future because I am going to spend the rest of my life there.
– Charles F. Kettering

Business as usual is dead. Green growth is the answer to both our climate and economic problems.
– Anders Fogh Rasmussen

Our generation has inherited an incredibly beautiful world from our parents and they from their parents. It is in our hands whether our children and their children inherit the same world. We must not be the generation responsible for irreversibly damaging the environment.
Richard Branson

ISBN: 978-0170210348

1 Write down your top five comments about sustainability and rank them in order, beginning with the one you like best.

2 Give a brief comment about what each saying means.

3 Either make a drawing or describe the drawing you would get, to go with each saying.

4 Explain why the human activities of war and terrorism are enemies of sustainability.

5 Find other sayings about sustainability. Choose your favourite.

6 In your group make up a short piece such as a PowerPoint presentation or poem to show how sustainability is a positive force in the world.

ISBN: 978-0170210348